I0045143

Peter W. Latham

The Croonian Lectures

On Some Points in the Pathology of Rheumatism, Gout and Diabetes

Peter W. Latham

The Croonian Lectures
On Some Points in the Pathology of Rheumatism, Gout and Diabetes

ISBN/EAN: 9783337162924

Printed in Europe, USA, Canada, Australia, Japan

Cover: Foto ©berggeist007 / pixelio.de

More available books at **www.hansebooks.com**

THE CROONIAN LECTURES

ON

SOME POINTS IN THE
PATHOLOGY OF RHEUMATISM, GOUT, AND
DIABETES.

DELIVERED AT THE ROYAL COLLEGE OF PHYSICIANS, LONDON,
MARCH 30; APRIL 1, 6, 1886.

BY

P. W. LATHAM, M.A., M.D., F.R.C.P.,

DOWNING PROFESSOR OF MEDICINE IN THE UNIVERSITY OF CAMBRIDGE;
SENIOR PHYSICIAN TO ADDENBROOKE'S HOSPITAL, CAMBRIDGE, AND FORMERLY
ASSISTANT PHYSICIAN TO THE WESTMINSTER HOSPITAL.

CAMBRIDGE:
DEIGHTON, BELL AND CO.
LONDON: G. BELL AND SONS.
1887

ON SOME POINTS IN THE PATHOLOGY OF RHEUMATISM, GOUT AND DIABETES.

LECTURE I.

MR PRESIDENT AND GENTLEMEN,—In rheumatism, gout and diabetes, we have three disorders about the pathology of which there is still much that is obscure and unsettled. On this ground alone, therefore, they present a very attractive and interesting field for investigation and study; and the interest is intensified by the conviction that a clearer insight, as regards the changes which take place in either of these disorders, would furnish the clue by which to unravel many of the most important phenomena which are associated with other diseases.

I class these three disorders together because they seem to possess a certain relationship with each other. Cases occur in which it is difficult to decide whether the disease must be regarded as rheumatic or as gouty; not infrequently transient diabetes appears as the harbinger of a gouty attack; and on the other hand, gouty, rheumatic or neuralgic pains are very common accompaniments of diabetes. In all we have changes, differing however in character, showing themselves in the blood; the result of abnormal metabolism either in the muscular or glandular tissues.

What are the changes which take place in the tissues, or in the blood? In attempting to answer these enquiries we meet at the very outset of our investigation with a most serious difficulty.

Before we can attempt to unravel the changes which may occur in any particular tissue or fluid either in health or disease we must know what is the constitution of the particular substance we are about to investigate. I do not mean its ultimate chemical composition—that is easily arrived at—but its proximate constituents, its structural formula. It seems almost superfluous to make such a statement. But the difficulty which presents itself is, that we do not know what are the proximate elements which make up living tissue; nor what are the chemical changes which take place as it performs its function; nor the alteration in the arrangement of its molecules as it passes from an active to an effete state. By chemical investigation it has been shown that various complex bodies which did not exist, as such, in the living tissue can be extracted from the different tissues after death; in the process of dying a molecular change of some kind takes place, the ultimate atoms are no longer arranged as they were in the living state, and new substances are formed. The chemical features of dead and living tissue are strikingly different from each other. Blood, for instance, when shed from the vessels of a living body is perfectly fluid—the moment it is shed it undergoes changes; in a short time coagulation takes place, and a certain substance fibrin appears in the blood plasma, the coagulation being the result of a chemical process between certain factors in the blood and the conversion of living *plasmine* or some part of plasmine (fibrinogen) into fibrin.

A living muscle possesses "irritability" which it loses in dying and is succeeded by what is termed 'rigor mortis,' during the onset of which chemical change takes place and the muscle being previously neutral or faintly alkaline acquires a distinctly acid reaction; and this change is accompanied by a large and sudden development of carbonic acid. From dead muscle by a certain process we can obtain myosin, one of the proteids, which does not exist however in that form in living muscle. From living muscle we can obtain a slightly opalescent substance termed *muscle plasma*, which at first is quite fluid, but when exposed at the ordinary temperature becomes a solid jelly splitting up afterwards into clot and acid serum. The loose granular and flocculent clot is myosin which has been produced by some chemical change in the living plasma. The serum contains other proteid substances, serum-albumin, &c., closely resembling myosin and fibrin in their properties and more especially in their chemical composition. Now notwithstanding numerous researches, no definite agreement has been arrived at as to the constitution of dead proteid or albumenoid substance, or as to the manner in which the various substances leucine, tyrosine, &c. are contained in it. The formula by which according to Lieberkühn a proteid or albumenoid may be approximately represented is

$$C_{72}H_{112}N_{18}O_{22}S$$

and this according to Hlasiwetz and Habermann by hydration may be split up into aspartic acid, glutamic acid, leucine, tyrosine and ammonia. Schützenberger obtained other products of decomposition but none of the attempts hitherto made to assign a molecular structure to the substance have been considered successful. If such is the case as regards dead tissue, it may appear somewhat presumptuous and rash on my part to endeavour to assign a molecular structure chemically to living tissue. But if this problem could be solved, if we were acquainted with the normal chemical changes which take place in the proximate constituents of a tissue in health and which when modified produced disease, what a flood of light would be thrown upon many points in pathology and therapeutics which at present are very dimly comprehended; and here I need only refer to the action of so-called alterative remedies such as iodine, bromine, mercury and arsenic. Even at the imminent risk of failure then, I think it is worth while to make some attempt to solve the problem, and to bring forward some of the points which have been suggested to me by the action of remedies in disease. I may at least advance a theory which to myself appears to explain many changes in the tissues; and I shall be content if, whilst endeavouring to arrive

at the truth, what I say may direct attention to the subject and
be useful to others more competent than myself to work out its
solution.

A certain number of substances which can be obtained from
albuminous material, or which are developed in the animal
economy can be produced in more or less diverse ways in the
laboratory—lactic acid, leucine, glycocine, &c.

In examining and investigating the various methods by which
these bodies may be prepared artificially in the laboratory, we
come upon the remarkable fact that a large number of them can
be obtained from a particular series of cyanogen compounds, the
so-called cyan-alcohols or cyanhydrins—bodies which may be
obtained by oxidising the various alcohols, and so forming the
aldehyde, and then combining this with hydrocyanic acid. For
instance,

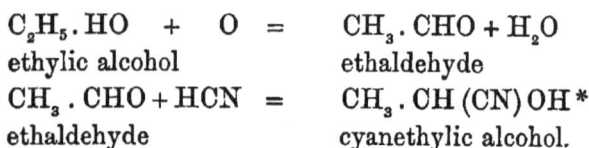

$$C_2H_5.HO + O = CH_3.CHO + H_2O$$
ethylic alcohol   ethaldehyde
$$CH_3.CHO + HCN = CH_3.CH(CN)OH*$$
ethaldehyde   cyanethylic alcohol.

Now these cyan-alcohols are very unstable bodies, readily
undergoing change. Treated with ammonia they form a series of
cyanamides which also are very unstable, and easily undergo
condensation, being converted into imido-nitriles with elimination
of ammonia. These facts at once suggest the enquiry: Have we
not in these cyanogen compounds substances possessing some properties that belong to living tissue, namely, those of undergoing
intra-molecular change, and also condensation? And, further, if
from these substances we can obtain the various products which
result from the disintegration of albumen, may not albumen itself
be simply a compound made up of these elements?

Let me indicate some of the animal products which may be
obtained from these cyan-alcohols, and show how these various
substances are formed.

* Miller's *Elements of Chemistry*, Part III., 1880, p. 736.

Leucine is very largely diffused in the animal organism, and may be obtained by various processes from albumen, flesh, gelatine, casein, &c.*

By oxidising amylic alcohol with potassium chromate and sulphuric acid, and distilling, we obtain amylic or valerianic aldehyde†—

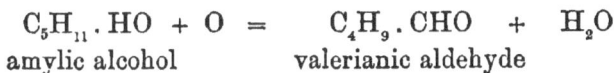

$$C_5H_{11} . HO + O = C_4H_9 . CHO + H_2O$$
amylic alcohol   valerianic aldehyde

Mixed with aqueous ammonia the aldehyde is converted into valeral ammonia, and this digested with hydrocyanic acid and hydrochloric acid is converted into leucine—

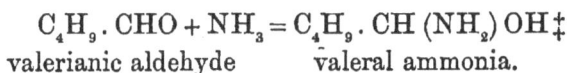

$$C_4H_9 . CHO + NH_3 = C_4H_9 . CH (NH_2) OH‡$$
valerianic aldehyde   valeral ammonia.

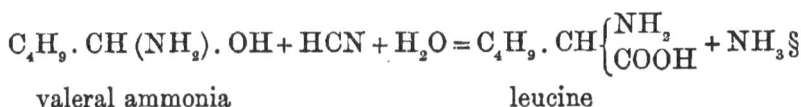

$$C_4H_9 . CH (NH_2) . OH + HCN + H_2O = C_4H_9 . CH\begin{cases} NH_2 \\ COOH \end{cases} + NH_3 §$$
valeral ammonia   leucine

This is the usual way of obtaining leucine artificially; but Tiemann has shown‖ that the amido-acids, both of the fatty and

* Watts, *Dictionary*, Vol. III. p. 574.
† *Ib.* Vol. v. p. 973.
‡ *Ib.* Vol. v. p. 974.
§ Fownes, *Manual of Org. Chem.*, 1877, p. 385.
‖ *Berichte der deutsch. chem. Gesell.*, XIV. s. 1985. " The amido acids of the fatty series are easily obtained by the familiar reactions which take place on treating aldehyde ammonia with hydrochloric and hydrocyanic acids, and which led Strecker to the discovery of alanine. . . . . The reactions indicated by Strecker take place unquestionably according to the following general formulæ:

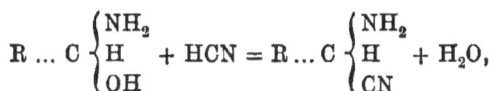

$$R ... C \begin{cases} NH_2 \\ H \\ OH \end{cases} + HCN = R ... C \begin{cases} NH_2 \\ H \\ CN \end{cases} + H_2O,$$

and

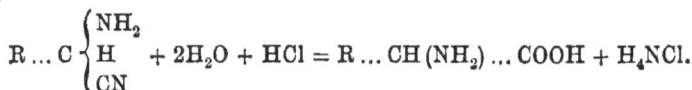

$$R ... C \begin{cases} NH_2 \\ H \\ CN \end{cases} + 2H_2O + HCl = R ... CH (NH_2) ... COOH + H_4NCl.$$

aromatic series, may be obtained by converting the aldehydes and ketones into cyan-alcohols, and then into amido-nitriles or cyan-amides. We may consequently also have the following changes:

$$C_4H_9 . CHO + HCN = C_4H_9 . CH {OH \atop CN}$$

valerianic aldehyde   pentene cyan-alcohol

$$C_4H_9 . CH{OH \atop CN} + NH_3 = C_4H_9 . CH{NH_2 \atop CN} \quad + \quad H_2O$$

pentene cyanamide

$$C_4H_9 . CH {NH_2 \atop CN} + 2H_2O = C_4H_9 . CH {NH_2 \atop COOH} + NH_3 .$$

leucine

Leucine prepared in this way is not quite identical with animal leucine. The explanation I venture to suggest is, that just as sarcolactic acid is a compound of ethidene and ethene lactic acids, so leucine is a compound of several amido-acids, the molecule C_5H_{10} being made up in different ways

$$C_3H_6 . CH_3 . CH ; (CH_2)_3CH_3 . CH ; \&c.,$$

or it may be prepared in the following manner:—

The question arises, whether the cyanamide

$$R ... C {NH_2 \atop H \atop CN}$$

could not be obtained more readily from the cyanhydrides of the aldehydes

$$R ... C {OH \atop H \atop CN}$$

by digesting them with ammonia, expecting the ultimate change to be as follows:

$$R ... C {OH \atop H \atop CN} + NH_3 = R ... C {NH_2 \atop H \atop CN} + H_2O.$$

The truth of this supposition has been confirmed by experiment."—*Berichte*, XIII. s. 382.

By the oxidation of the pentyl alcohol, diethyl carbinol, we obtain diethyl ketone[*]:

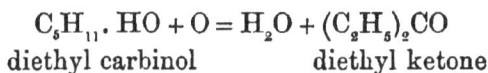

$$C_5H_{11} \cdot HO + O = H_2O + (C_2H_5)_2CO$$

diethyl carbinol   diethyl ketone

Tiemann[†] has shown that this with hydrocyanic acid is converted into a cyanhydrin, which, acted upon by ammonia and then by an acid, produces leucine—

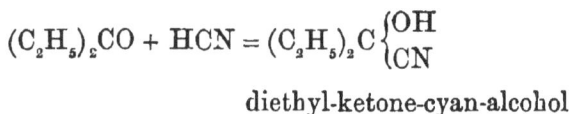

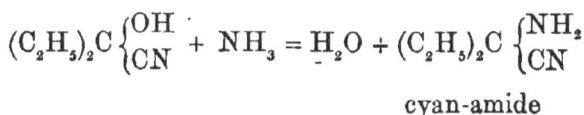

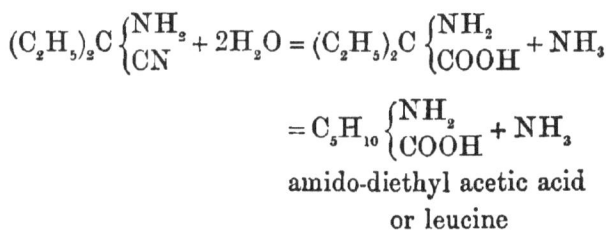

$$(C_2H_5)_2CO + HCN = (C_2H_5)_2C \begin{cases} OH \\ CN \end{cases}$$

diethyl-ketone-cyan-alcohol

$$(C_2H_5)_2C \begin{cases} OH \\ CN \end{cases} + NH_3 = H_2O + (C_2H_5)_2C \begin{cases} NH_2 \\ CN \end{cases}$$

cyan-amide

$$(C_2H_5)_2C \begin{cases} NH_2 \\ CN \end{cases} + 2H_2O = (C_2H_5)_2C \begin{cases} NH_2 \\ COOH \end{cases} + NH_3$$

$$= C_5H_{10} \begin{cases} NH_2 \\ COOH \end{cases} + NH_3$$

amido-diethyl acetic acid
or leucine

Butene cyan-alcohol $C_4H_8 \begin{cases} OH \\ CN \end{cases}$—may in like manner be prepared from the butyric aldehydes, from aldol or butene glycol; and from one form of this we obtain Amido-isovaleric acid, which occurs in the pancreas of the ox[‡].

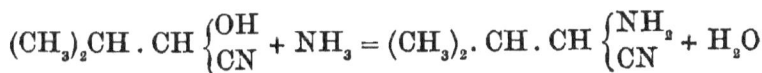

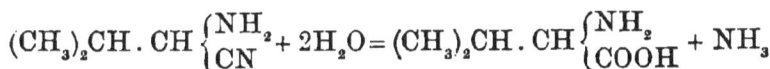

$$(CH_3)_2CH \cdot CH \begin{cases} OH \\ CN \end{cases} + NH_3 = (CH_3)_2 \cdot CH \cdot CH \begin{cases} NH_2 \\ CN \end{cases} + H_2O$$

$$(CH_3)_2CH \cdot CH \begin{cases} NH_2 \\ CN \end{cases} + 2H_2O = (CH_3)_2CH \cdot CH \begin{cases} NH_2 \\ COOH \end{cases} + NH_3$$

amido-isovaleric acid

[*] Fownes, Manual of Org. Chem., 1877, p. 153.
[†] Berichte, xiv. s. 1975.   [‡] Fownes, p. 385.

Amido-valeric acid or butalanine was also obtained by Schützenberger from albumen*.

Similarly by the oxidation of propyl alcohol, propionic aldehyde is obtained, which may be converted into propene cyan-
alcohol †—

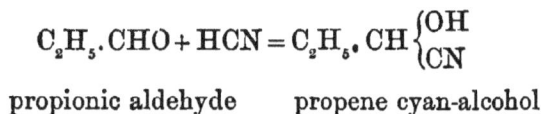

$$C_2H_5.CHO + HCN = C_2H_5.CH \begin{cases} OH \\ CN \end{cases}$$

propionic aldehyde   propene cyan-alcohol

By oxidation of isopropyl alcohol, dimethyl ketone or acetone
may be produced, and then converted into dimethyl-ketone-cyan-
alcohol‡—

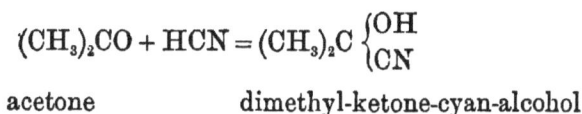

$$(CH_3)_2CO + HCN = (CH_3)_2C \begin{cases} OH \\ CN \end{cases}$$

acetone   dimethyl-ketone-cyan-alcohol

Propene cyan-alcohol may also be obtained, by converting
propene glycol into chlor-alcohol§ and acting upon this with potassium cyanide—

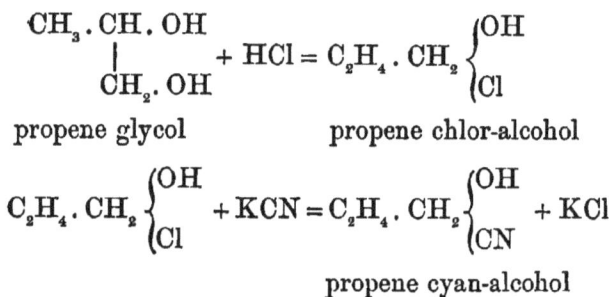

$$\begin{array}{c} CH_3.CH.OH \\ | \\ CH_2.OH \end{array} + HCl = C_2H_4.CH_2 \begin{cases} OH \\ Cl \end{cases}$$

propene glycol   propene chlor-alcohol

$$C_2H_4.CH_2 \begin{cases} OH \\ Cl \end{cases} + KCN = C_2H_4.CH_2 \begin{cases} OH \\ CN \end{cases} + KCl$$

propene cyan-alcohol

Each of these cyan-alcohols, treated with ammonia and then
with acids, will yield the corresponding amido-butyric acid, a substance also obtained by Schützenberger from albumen—

* *Comptes Rendus*, t. 81, p. 1110.
† Fownes, *Manual of Org. Chem.*, 1877, p. 329.
‡ *Ibid.* p. 329.   § *Ibid.* p. 176.

$$C_3H_6 \begin{cases} OH \\ CN \end{cases} + NH_3 = C_3H_6 \begin{cases} NH_2 \\ CN \end{cases} + H_2O$$

$$C_3H_6 \begin{cases} NH_2 \\ CN \end{cases} + 2H_2O = C_3H_6 \begin{cases} NH_2 \\ COOH \end{cases} + NH_3$$

amido-butyric acid

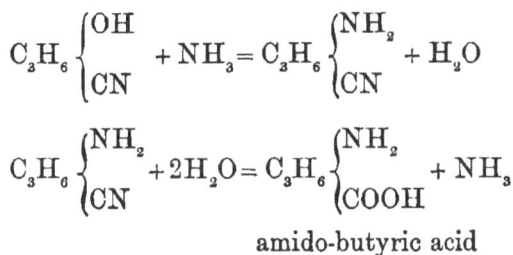

If the cyan-alcohols are treated directly with acids we obtain the corresponding oxybutyric acid—

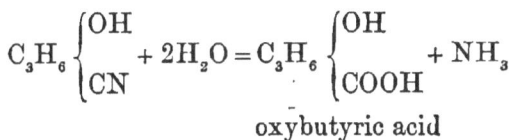

$$C_3H_6 \begin{cases} OH \\ CN \end{cases} + 2H_2O = C_3H_6 \begin{cases} OH \\ COOH \end{cases} + NH_3$$

oxybutyric acid

By oxidation oxy-isobutyric acid is resolved into carbonic acid and acetone, a substance which sometimes appears in diabetic urine—

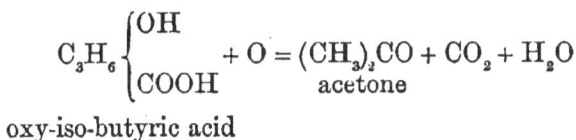

$$C_3H_6 \begin{cases} OH \\ COOH \end{cases} + O = (CH_3)_2CO + CO_2 + H_2O$$

oxy-iso-butyric acid $\qquad$ acetone

Ethidene and ethene cyan-alcohols may be prepared (i) by oxidising ethylic alcohol, and treating the aldehyde so obtained with hydrocyanic acid —

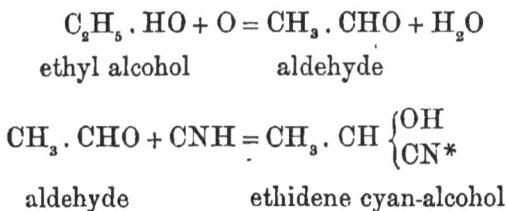

$$C_2H_5 . HO + O = CH_3 . CHO + H_2O$$

ethyl alcohol $\qquad$ aldehyde

$$CH_3 . CHO + CNH = CH_3 . CH \begin{cases} OH \\ CN* \end{cases}$$

aldehyde $\qquad$ ethidene cyan-alcohol

(ii) By converting ethene alcohol or glycol into a cyan-alcohol—

---

* Fownes, *Manual*, p. 319.

$$\begin{array}{l} CH_2 . OH \\ | \qquad\qquad + HCl = H_2O + CH_2 CH_2 \begin{cases} OH \\ Cl \end{cases} \\ CH_2 . OH \end{array}$$

glycol   chlorhydrin

$$CH_2 . CH_2 \begin{cases} OH \\ Cl \end{cases} + KCN = KCl + CH_2 . CH_2 \begin{cases} OH \\ CN* \end{cases}$$

ethene cyan-alcohol

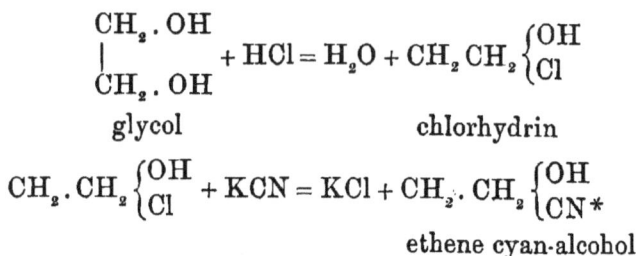

And from these cyan-alcohols the corresponding amido-acids or alanines may be obtained, viz.

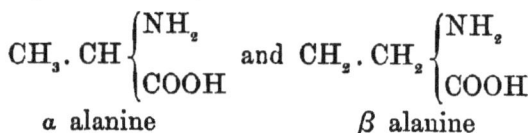

$$CH_3 . CH \begin{cases} NH_2 \\ COOH \end{cases} \quad \text{and} \quad CH_2 . CH_2 \begin{cases} NH_2 \\ COOH \end{cases}$$

a alanine   β alanine

Alanine or amido propronic acid was also among the products Schützenberger obtained from albumen.

Lactic acid, again, is an important product of the animal organism. It is developed when a living muscle contracts, and it is produced when a muscle dies. The variety of this acid which is obtained by the disintegration of albuminous compounds and is formed during the contraction or tetanus of muscular fibres, and hence called sarco-lactic acid, may be regarded as a mixture of two kinds of lactic acid† the more abundant being paralactic acid or ethidene lactic acid $CH_3 . CH \begin{cases} OH \\ COOH \end{cases}$, the other ethene lactic acid $CH_2 . CH_2 \begin{cases} OH \\ COOH \end{cases}$. Now (i) by treating ethidene cyan-alcohol with acids or alkalies paralactic or ethidene lactic acid is obtained

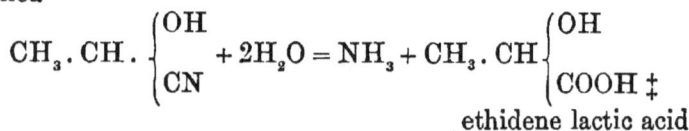

$$CH_3 . CH . \begin{cases} OH \\ CN \end{cases} + 2H_2O = NH_3 + CH_3 . CH \begin{cases} OH \\ COOH \ddagger \end{cases}$$

ethidene lactic acid

* Fownes, *Manual*, p. 319.
† Watts, *Dictionary of Chemistry*, Vol. VIII. p. 1160.
‡ Fownes, *Manual*, p. 319.

(ii) by treating ethene cyan-alcohol in the same way ethene lactic acid is obtained.

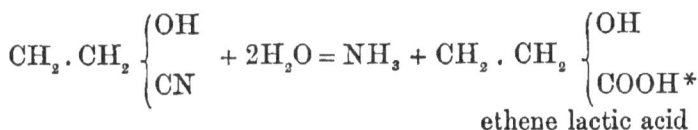

$$CH_2 . CH_2 \begin{cases} OH \\ CN \end{cases} + 2H_2O = NH_3 + CH_2 . CH_2 \begin{cases} OH \\ COOH* \end{cases}$$

ethene lactic acid

Methene cyan-alcohol, $CH_2 \begin{cases} OH \\ CN \end{cases}$. Though methylic aldehyde has not yet been isolated, but only obtained in solution in methylic alcohol†, we may infer from the reactions of the other members of the alcoholic series that similar changes will be produced here. That by oxidising methylic alcohol the aldehyde will be obtained, which combined with hydrocyanic acid will form methene cyan-alcohol

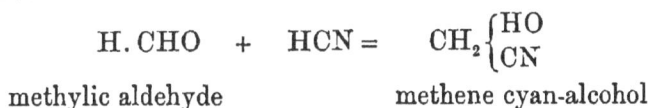

$$H . CHO \quad + \quad HCN = \quad CH_2 \begin{cases} HO \\ CN \end{cases}$$

methylic aldehyde   methene cyan-alcohol

and from this we should obtain amido-acetic acid, or glycocine—

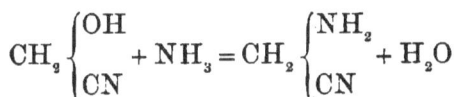

$$CH_2 \begin{cases} OH \\ CN \end{cases} + NH_3 = CH_2 \begin{cases} NH_2 \\ CN \end{cases} + H_2O$$

and

$$CH_2 \begin{cases} NH_2 \\ CN \end{cases} + 2H_2O = CH_2 \begin{cases} NH_2 \\ COOH \end{cases} + NH_3$$

glycocine

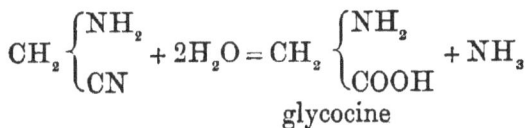

Again, by acting on albumenoids with certain oxidising agents such as a mixture of potassic dichromate or manganic dioxide with sulphuric acid, Guckelberger‡ obtained the following products:—

Caproic acid,	Propionic acid,
Valeric ,,	Acetic ,,
Butyric ,,	Formic ,,

* Fownes, *Manual*, p. 319.
† Miller's *Elements of Chemistry*, Part III., 1880, p. 723.
‡ Leibig's *Annal.*, lxiv. 39.

Now all these acids can be obtained from the bodies under discussion, viz. the cyan-alcohols. By treating them with acids we have first of all the corresponding oxyacid formed.

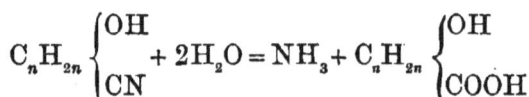

$$C_nH_{2n}\begin{cases}OH\\CN\end{cases} + 2H_2O = NH_3 + C_nH_{2n}\begin{cases}OH\\COOH\end{cases}$$

and oxyacids of the form

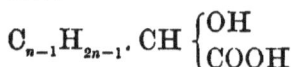

$$C_{n-1}H_{2n-1}.\,CH\begin{cases}OH\\COOH\end{cases}$$

when oxidised are converted into CO_2, the corresponding acid of the acetic series and water,

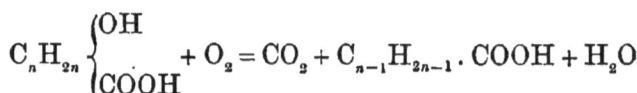

$$C_nH_{2n}\begin{cases}OH\\COOH\end{cases} + O_2 = CO_2 + C_{n-1}H_{2n-1}.\,COOH + H_2O$$

ethidene cyan-alcohol $CH_3.\,CH\begin{cases}OH\\CN\end{cases}$, for instance, may be converted into lactic acid $CH_3.\,CH\begin{cases}OH\\COOH\end{cases}$, which treated with chromic acid mixture is oxidised to formic and acetic acids[*]—

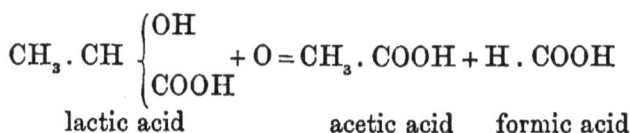

$$CH_3.\,CH\begin{cases}OH\\COOH\end{cases} + O = CH_3.\,COOH + H.\,COOH$$
  lactic acid   acetic acid   formic acid

and the formic acid may be further oxidised into carbonic acid and water.

In the same way butene cyan-alcohol is converted into oxy-valeric acid $C_3H_7.\,CH\begin{cases}OH\\COOH\end{cases}$ which oxidised is transformed into carbonic dioxide and butyric acid[†].

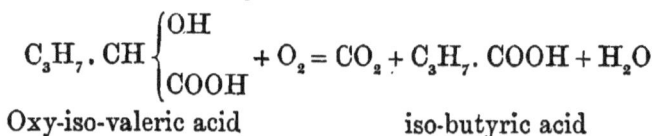

$$C_3H_7.\,CH\begin{cases}OH\\COOH\end{cases} + O_2 = CO_2 + C_3H_7.\,COOH + H_2O$$
  Oxy-iso-valeric acid   iso-butyric acid

[*] Fownes, p. 325.   [†] *Ibid.* p. 330.

From these cyan-alcohols then, we can obtain the corresponding amido-acids, glycocine leucine, &c., and all the acids of the acetic series, as well as those of the lactic acid series. This being the case we come to the question, Can we reverse the process, as we know can be done chemically in many similar cases? Can the amido-acids be converted into the cyan-alcohols, and can we then theoretically build up albumen from such constituents?

The first step towards this is, I think, indicated in what follows when amylaceous matter or sugar is introduced into the alimentary canal. A certain accumulation of glycogen takes place in the liver, and further, a considerable quantity of sugar can be slowly injected into the portal vein without any appearing in the urine. The sugar, therefore, as it passes into the liver loses a molecule of H_2O and glycogen is formed. We cannot as yet tell how glucose in a living plant is converted into starch, nor can we tell in what way the liver transforms the glucose into glycogen, all we know is that in some way in the tissue glucose is dehydrated and glycogen is formed*

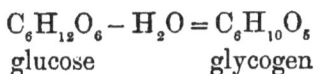

$$C_6H_{12}O_6 - H_2O = C_6H_{10}O_5$$
glucose   glycogen

Now in order to obtain the various amido-acids from albuminous tissue, the reverse process must be adopted; the tissue must be hydrated, that is, a certain amount of water must enter into combination with it†. Some of these amido bodies are found in the alimentary canal; glycocine and taurine appear as the result of the decomposition in the duodenum of the biliary acids, glyco- and tauro-cholic acids, these two acids by hydration splitting up into cholic acid and glycocine, and cholic acid and taurine respectively.

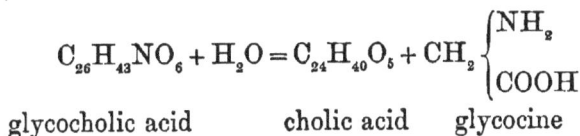

$$C_{26}H_{43}NO_6 + H_2O = C_{24}H_{40}O_5 + CH_2 \begin{cases} NH_2 \\ COOH \end{cases}$$
glycocholic acid   cholic acid   glycocine

* See Foster's *Physiology*, 4th ed. pp. 418—422.
† See Schützenberger's experiments, *Comptes Rendus*, ts. 80, 81, 84.

$$C_{26}H_{45}NSO_7 + H_2O = C_{24}H_{40}O_5 + C_2H_4 \begin{cases} NH_2 \\ SO_3H \end{cases}$$

taurocholic acid   cholic acid   taurine

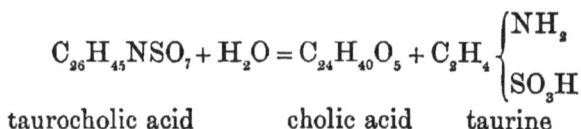

the cholic acid passing off by the intestines, the glycocine and taurine being reabsorbed.

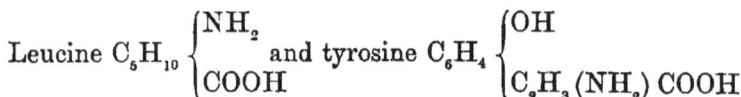

$$\text{Leucine } C_5H_{10} \begin{cases} NH_2 \\ COOH \end{cases} \text{and tyrosine } C_6H_4 \begin{cases} OH \\ C_2H_3 (NH_2) COOH \end{cases}$$

are also among the substances produced when proteid substances are digested in the alimentary canal. What becomes of them?

" One result of the action of the pancreatic juice is the formation of considerable quantities of leucine and tyrosine. In dealing with the statistics of nutrition, our attention will be drawn to the fact that the introduction of proteid matter into the alimentary canal is followed by a large and rapid excretion of urea, suggesting the idea that a certain part of the total quantity of the urea normally secreted comes from a direct metabolism of the proteids of the food, without these really forming a part of the tissues of the body. We do not know to what extent normal pancreatic digestion has for its product leucine, and its companion tyrosine; but if, especially when a meal rich in proteids has been taken, a considerable quantity of leucine is formed, we can perceive an easy and direct source of urea, provided that the metabolism of the body is capable of converting leucine into urea. That the body can effect this change is shown by the fact that leucine, when introduced into the alimentary canal in even large quantities, does reappear in the urine as urea; that is, the urine contains no leucine, but its urea is proportionately increased; and the same is probably the case with tyrosine. Now the leucine formed in the alimentary canal is carried by the portal blood straight to the liver; and the liver, unlike other glandular organs, does, even in a perfectly normal state of things, contain urea We are thus led to the view that among the numerous metabolic

events which occur in the hepatic cells, the formation of urea out of leucine or out of other antecedents may be ranked as one......
Probable, therefore, as this view may seem, it has not as yet been established as a fact."

"The view that leucine is transformed into urea lands us, however, in very considerable difficulties. Leucine, as we know, is amido-caproic acid; and, with our present chemical knowledge, we can conceive of no other way in which leucine can be converted into urea than by the complete reduction of the former to the ammonia condition (the caproic acid residue being either elaborated into a fat or oxidised into carbonic acid) and by a reconstruction of the latter out of the ammonia so formed. We have a somewhat parallel case in glycin (glycocine). This, which is amido-acetic acid, when introduced into the alimentary canal also reappears as urea; here too, a reconstruction of urea out of an ammonia phase must take place*."

Again,

"To ascertain the influence of the liver in the formation of urea, Solnikoff has established a direct connexion between the portal and jugular veins by means of an india-rubber tube, an operation which, if carefully performed, is borne with impunity. Vascular pressure at first lowered, soon returns to the normal. The urinary secretion was, however, completely arrested, and was not re-established until urea had been injected into the veins of the animal. The only effect of these injections was to raise the vascular pressure. The results were in no way modified by the preceding section of the splanchnic nerves. If on the other hand, instead of the portal, the crural vein is placed in communication with the jugular vein, the urinary secretion is unchanged. It is thus clear that the urea which passes out of the system by the kidneys, enters the circulation with the blood which issues from the liver †."

* Foster's *Physiology*, 4th ed. pp. 438, 439.
† *Lancet*, Dec. 3, 1881, p. 971.

So much for the physiological side. Let us turn to the pathological. Frerichs* pointed out that in acute atrophy of the liver, the urine contains a large quantity of leucine and tyrosine, and that urea almost or entirely disappears.

It is evident from these statements that these amido-acids, when introduced into the alimentary canal of a healthy animal, are carried by the portal blood into the liver and there undergo some change, urea being one of the products. It is hardly intelligible, however, that the nitrogen in the fresh nutritive material should immediately be transformed into urea, and as such at once pass out of the system; and I shall endeavour presently to show that this is not the case, and subsequently to show that some of the glycocine, at least, is not concerned in the formation of urea.

These amido-bodies or glycines are capable of uniting with each other (Hofmeister) and it is probable that their molecular weights are at least double as great as their formulæ would indicate†. They may be represented by the formula

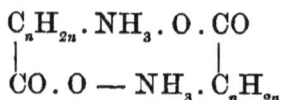

$$C_nH_{2n}.NH_3.O.CO$$
$$CO.O - NH_3.C_nH_{2n}$$

Now as the function of the liver in the formation of glycogen seems to be the dehydration of glucose, the question naturally suggests itself : What compounds would result from the liver acting in the same way on these duplex or triple glycines, what substances would be produced by their dehydration ? Can they in any way be converted into a series of cyanalcohols ?

From two molecules of glycocine dehydrated‡ we should have

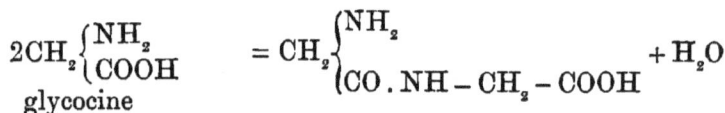

$$2CH_2 \begin{cases} NH_2 \\ COOH \end{cases} = CH_2 \begin{cases} NH_2 \\ CO.NH - CH_2 - COOH \end{cases} + H_2O$$

glycocine

---

* *Klinik der Leberkrankheiten*, 1858, s. 206.

† Miller's *Chemistry*, 1880, Part III. p. 866.

‡ In dealing with the formation of uric acid later on I shall show that dehydration of glycocine does take place in the living body in the formation of hydantoin, see page 67.

A very interesting point arises here. In 1828 Wöhler showed that urea $CO \begin{Bmatrix} NH_2 \\ NH_2 \end{Bmatrix}$ can be produced by molecular transformation of ammonium cyanate $CNO.NH_4$, this change taking place simply by heating a solution of the latter. Pflüger* in calling attention to the great molecular energy of the cyanogen compounds suggested that the functional metabolism of protoplasm by which energy is set free, may be compared to the conversion of the energetic unstable cyanogen compounds into the less energetic and more stable amides. In other words, that ammonium cyanate is a type of living, and urea of dead nitrogen, and the conversion of the former into the latter is an image of the essential change which takes place when a living proteid dies†.

In accordance with this view then $CO.NH$ in the above formula for the dehydration of glycocine would represent dead nitrogen and as the substance becomes part of a living tissue it would be transformed into $CNOH$.

The two molecules of glycocine therefore on passing through the liver would be transformed as follows :

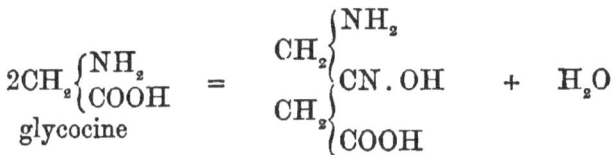

$$2CH_2 \begin{Bmatrix} NH_2 \\ COOH \end{Bmatrix} = \begin{matrix} CH_2 \\ \\ CH_2 \end{matrix} \begin{Bmatrix} NH_2 \\ CN.OH \\ \\ COOH \end{Bmatrix} + H_2O$$
glycocine

Similarly

$$3CH_2 \begin{Bmatrix} NH_2 \\ COOH \end{Bmatrix} = \begin{matrix} CH_2 \\ \\ CH_2 \\ \\ CH_2 \end{matrix} \begin{Bmatrix} NH_2 \\ CN.OH \\ CN.OH \\ \\ COOH \end{Bmatrix} + 2H_2O$$
glycocine

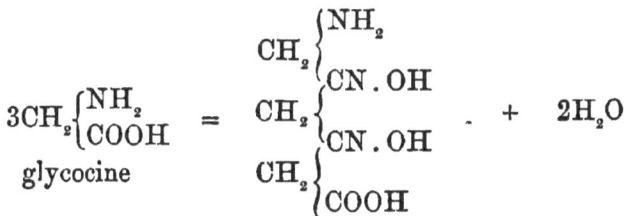

* Pflüger's *Archiv*, Bd. x. s. 337.
† Foster's *Physiology*, 4th ed. p. 749.

$$3C_2H_4\begin{cases}NH_2\\COOH\end{cases} = C_2H_4\begin{cases}C_2H_4\begin{cases}NH_2\\CN.OH\end{cases}\\CN.OH\\C_2H_4\begin{cases}\\COOH\end{cases}\end{cases} + 2H_2O$$

alanine

$$\text{and } 3C_5H_{10}\begin{cases}NH_2\\COOH\end{cases} = C_5H_{10}\begin{cases}C_5H_{10}\begin{cases}NH_2\\CN.OH\end{cases}\\CN.OH\\C_5H_{10}\begin{cases}\\COOH\end{cases}\end{cases} + 2H_2O$$

leucine

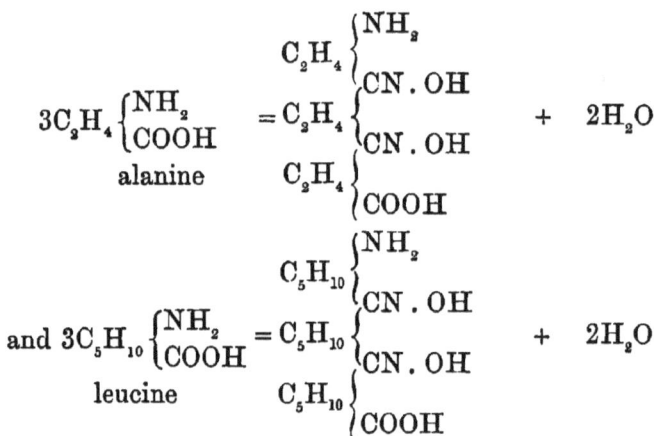

and in the same way these compound molecules may be connected together by combining the COOH in one with the NH_2 in the other, with elimination of H_2O.

By dehydration of the glycines then we should have generally :

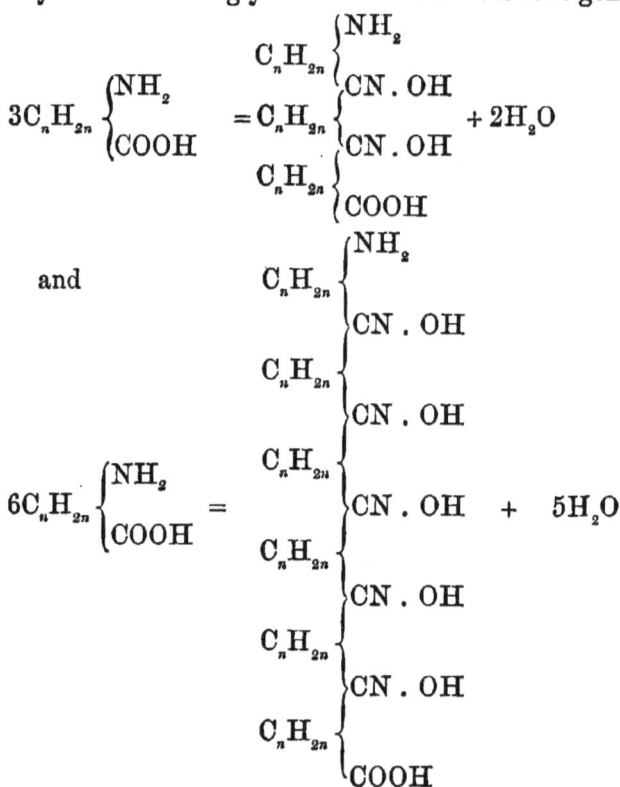

$$3C_nH_{2n}\begin{cases}NH_2\\COOH\end{cases} = C_nH_{2n}\begin{cases}C_nH_{2n}\begin{cases}NH_2\\CN.OH\end{cases}\\CN.OH\\C_nH_{2n}\begin{cases}\\COOH\end{cases}\end{cases} + 2H_2O$$

and

$$6C_nH_{2n}\begin{cases}NH_2\\COOH\end{cases} = C_nH_{2n}\begin{cases}NH_2\\\\CN.OH\end{cases}\begin{cases}\\CN.OH\end{cases} + 5H_2O$$

$$C_nH_{2n}\begin{cases}NH_2\\CN.OH\end{cases}$$
$$C_nH_{2n}\begin{cases}CN.OH\end{cases}$$
$$C_nH_{2n}\begin{cases}CN.OH\end{cases}$$
$$C_nH_{2n}\begin{cases}CN.OH\end{cases}$$
$$C_nH_{2n}\begin{cases}CN.OH\end{cases}$$
$$C_nH_{2n}\begin{cases}\\COOH\end{cases}$$

and in this way we arrive at an *atomic* combination of the
cyan-alcohols $C_nH_{2n}\begin{cases}OH\\CN\end{cases}$ united to a cyanamide, $C_nH_{2n}\begin{cases}NH_2\\CN\end{cases}$
and an acid $C_nH_{2n}\begin{cases}OH\\COOH\end{cases}$.

Another of the glycines, viz. tyrosine $C_6H_4\begin{cases}OH\\C_2H_3(NH_2)COOH\end{cases}$
or para-oxyphenyl-amido-propionic acid, which is obtained from
albumen and other proteid substances, may be thus represented

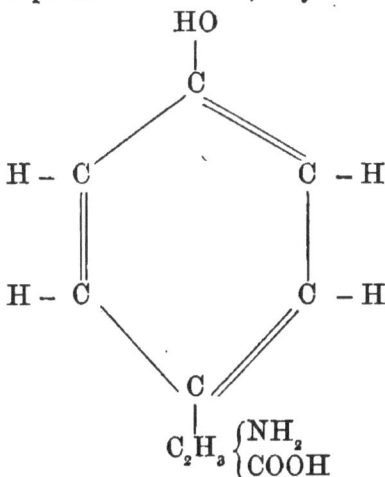

$$
\begin{array}{c}
HO\\
|\\
C\\
H-C \qquad\qquad C-H\\
\|\qquad\qquad\qquad\|\\
H-C \qquad\qquad C-H\\
C\\
|\\
C_2H_3\begin{cases}NH_2\\COOH\end{cases}
\end{array}
$$

and combining this with glycocine, H_2O being eliminated, we
should have

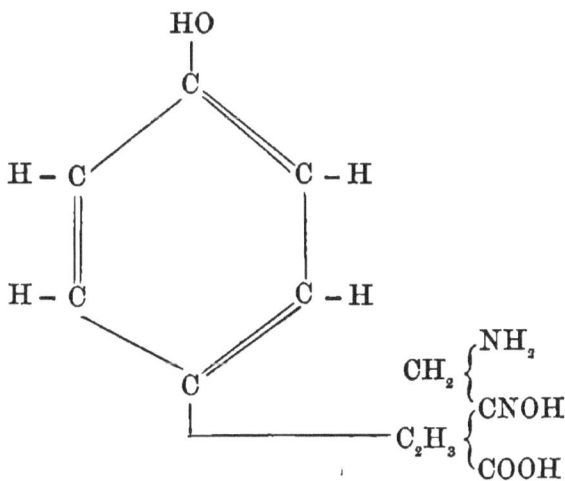

$$
\begin{array}{c}
HO\\
|\\
C\\
H-C \qquad\qquad C-H\\
\|\qquad\qquad\qquad\|\\
H-C \qquad\qquad C-H\\
C\\
|\qquad\qquad CH_2\begin{cases}NH_2\\ \\ \end{cases}\\
\quad\qquad C_2H_3\begin{cases}CNOH\\ \\COOH\end{cases}
\end{array}
$$

If now, bearing in mind that some compounds of cyanogen by condensation of three molecules form new compounds, such as

$$3CNOH = C_3N_3O_3H_3$$

cyanic acid   cyanuric acid

we combine this compound of tyrosine and glycocine (which as I shall presently try to show is itself made up of three molecules) with three molecules derived from glycocine

$$
CH_2 \begin{cases} NH_2 \end{cases}
$$
$$
CH_2 \begin{cases} CNOH \end{cases}
$$
$$
CH_2 \begin{cases} CNOH \end{cases}
$$
$$
CH_2 \begin{cases} COOH \end{cases}
$$

and three molecules derived from each of the other glycines in the series including leucine,

$$
C_2H_4 \begin{cases} NH_2 \\ CNOH \end{cases}
C_2H_4 \begin{cases} CNOH \end{cases}
C_2H_4 \begin{cases} COOH \end{cases}
\quad
C_3H_6 \begin{cases} NH_2 \\ CNOH \end{cases}
C_3H_6 \begin{cases} CNOH \end{cases}
C_3H_6 \begin{cases} COOH \end{cases}
\quad
C_4H_8 \begin{cases} NH_2 \\ CNOH \end{cases}
C_4H_8 \begin{cases} CNOH \end{cases}
C_4H_8 \begin{cases} COOH \end{cases}
\quad
C_5H_{10} \begin{cases} NH_2 \\ CNOH \end{cases}
C_5H_{10} \begin{cases} CNOH \end{cases}
C_5H_{10} \begin{cases} COOH \end{cases}
$$

these being combined together, the molecule COOH in the one series combining with NH_2 of the other, to form CNOH, with elimination of H_2O, we should have the compound

$$
\begin{array}{l}
C_5H_{10} \begin{cases} NH_2 \\ CNOH \end{cases} \\
2C_5H_{10} \begin{cases} CNOH \end{cases} \\
3C_4H_8 \begin{cases} CNOH \end{cases} \\
3C_3H_6 \begin{cases} CNOH \end{cases} \\
3C_2H_4 \begin{cases} CNOH \end{cases} \\
4CH_2 \begin{cases} CNOH \end{cases} \\
C_2H_3 \begin{cases} COOH \end{cases}
\end{array}
$$

(Benzene ring structure with HO–C at top, H–C and C–H on left and right, C at bottom connected by a line to C_2H_3)

the larger figures representing the number of times that cyan-alcohol is repeated; and combining the COOH and NH_2 at the ends of the chain with other molecules of the same composition, and replacing one of the hydrogen atoms of the benzene nucleus by SO_3H, as suggested by the composition of taurine, in which sulphur exists in the form of a sulphite, we should have the compound

$$
\begin{array}{c}
\text{HO} \\
|
\end{array}
$$

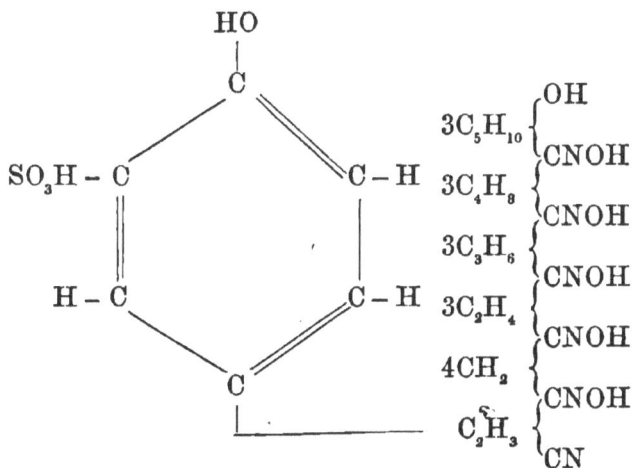

a body whose composition would be $C_{71}H_{117}N_{17}O_{21}S$; to which if the molecule CNOH were added we should have $C_{72}H_{118}N_{18}O_{22}S$, a formula almost identical with that given by Lieberkühn for albumen, viz. $C_{72}H_{112}N_{18}O_{22}S$. I will show further on how the addition is to be made.

Let me assume for the present that albumen is built up in this way of cyan-alcohols united to a benzene nucleus, the molecules held together by some force vital or otherwise, and consider whether a body so constituted would afford an explanation of the changes which take place in the muscular tissue in its active and quiescent state.

If the cyan-alcohols were detached separately they could, as I have already shown, be converted by hydration into the acids of the lactic acid series; by combining them with ammonia they could be converted into the various cyanamides, which hydrated are transformed into the glycines: glycocine, leucine, &c.

The acids when oxidised would be converted into their respective aldehydes, carbonic acid and water,

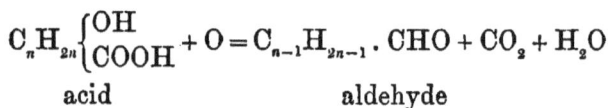

$$C_nH_{2n}\begin{cases}OH\\COOH\end{cases} + O = C_{n-1}H_{2n-1} . CHO + CO_2 + H_2O$$

acid   aldehyde

the aldehyde being then either combined with a fresh molecule of HCN to form a cyan-alcohol

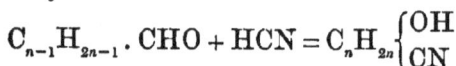

$$C_{n-1}H_{2n-1} . CHO + HCN = C_nH_{2n}\begin{cases}OH\\CN\end{cases}$$

and so again take its place as a constituent of albumen, or be further oxidised into CO_2 and H_2O

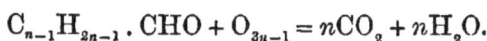

$$C_{n-1}H_{2n-1} . CHO + O_{3n-1} = nCO_2 + nH_2O.$$

Or the aldehyde from lactic acid may combine with the SO_3H disengaged from the benzene nucleus forming $C_2H_4\begin{cases}OH\\SO_3H\end{cases}$ which combining with ammonia produces $C_2H_4\begin{cases}OH\\SO_3NH_4\end{cases}$, and this in the laboratory may be transformed into Taurine $C_2H_4\begin{cases}NH_2\\SO_3H\end{cases}.$

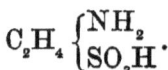

The glycines or amido-acids by oxidation may be converted into the nitriles HCN, $CH_3 . CN$, &c.

By the oxidation of glycocine for instance, we get

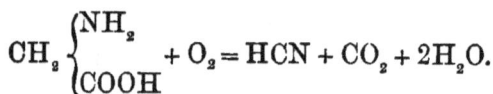

$$CH_2\begin{cases}NH_2\\COOH\end{cases} + O_2 = HCN + CO_2 + 2H_2O.$$

Leucine gives

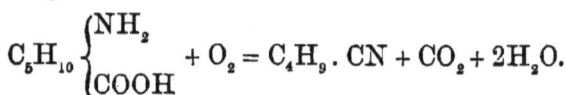

$$C_5H_{10}\begin{cases}NH_2\\COOH\end{cases} + O_2 = C_4H_9 . CN + CO_2 + 2H_2O.$$

By oxidising asparagin, we get

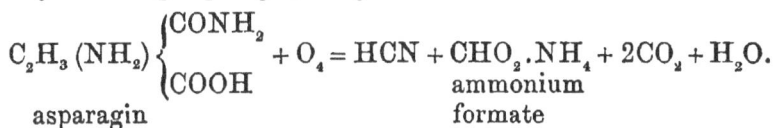

$$C_2H_3(NH_2)\begin{cases}CONH_2\\COOH\end{cases} + O_4 = HCN + CHO_2.NH_4 + 2CO_2 + H_2O.$$

asparagin   ammonium
                                                   formate

and ammonium formate dehydrated is converted into hydrocyanic acid.

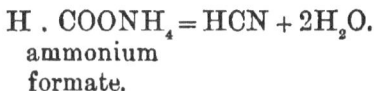

$$H.COONH_4 = HCN + 2H_2O.$$

ammonium
formate.

Now the molecules making up these cyan-alcohols may be detached in groups varying with the power which holds them together. The following is one change which may take place. By the hydration or decomposition of the separate cyan-alcohols lowest in the series we should have ammonia liberated, and the corresponding acid formed, glycollic acid, lactic acid &c.

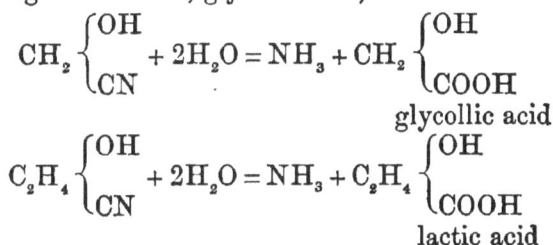

$$CH_2\begin{cases}OH\\CN\end{cases} + 2H_2O = NH_3 + CH_2\begin{cases}OH\\COOH\end{cases}$$

glycollic acid

$$C_2H_4\begin{cases}OH\\CN\end{cases} + 2H_2O = NH_3 + C_2H_4\begin{cases}OH\\COOH\end{cases}$$

lactic acid

The nascent ammonia may however combine with the hydroxyl of the next molecule in the chain to form a cyanamide, which hydrated would form the amido-acid, ammonia being again liberated and forming a cyanamide with the next molecule higher up; and so on all through the different series. The changes then may be thus represented

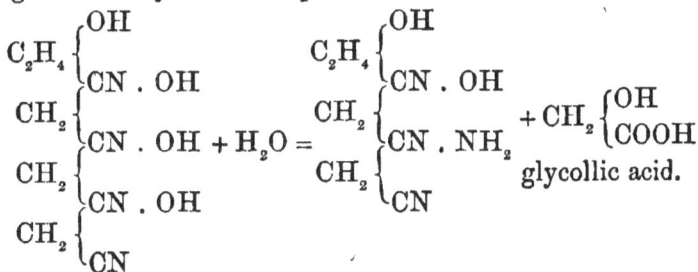

$$\begin{array}{l}C_2H_4\begin{cases}OH\\CN.OH\end{cases}\\CH_2\begin{cases}\\CN.OH\end{cases}+H_2O=\\CH_2\begin{cases}\\CN.OH\end{cases}\\CH_2\begin{cases}\\CN\end{cases}\end{array} \quad \begin{array}{l}C_2H_4\begin{cases}OH\\CN.OH\end{cases}\\CH_2\begin{cases}\\CN.NH_2\end{cases}\\CH_2\begin{cases}\\CN\end{cases}\end{array} + CH_2\begin{cases}OH\\COOH\end{cases}$$

glycollic acid.

Two different conditions now present themselves; (i) the molecule $CH_2 \begin{cases} NH_2 \\ CN \end{cases}$ may become detached, or (ii) the larger

molecule $\begin{matrix} CH_2 \\ CH_2 \end{matrix} \begin{cases} OH \\ CN \text{ . } NH_2 \\ CN \end{cases}$ may be separated from the chain. In

case (i) we should have by hydration

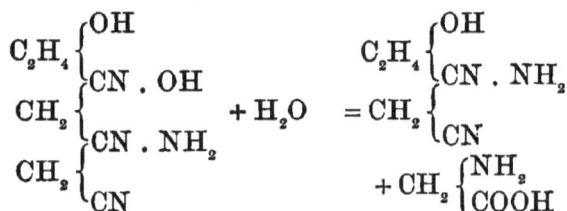

$$C_2H_4 \begin{cases} OH \\ CN \text{ . OH} \end{cases} \quad C_2H_4 \begin{cases} OH \\ CN \text{ . } NH_2 \end{cases}$$
$$CH_2 \begin{cases} CN \text{ . } NH_2 \end{cases} + H_2O \; = CH_2 \begin{cases} CN \end{cases}$$
$$CH_2 \begin{cases} CN \end{cases} \qquad\qquad + CH_2 \begin{cases} NH_2 \\ COOH \end{cases}$$

and so on, the remaining two molecules by hydration being converted into $CH_2 \begin{cases} NH_2 \\ COOH \end{cases}$ glycocine, $C_2H_4 \begin{cases} NH_2 \\ COOH \end{cases}$ alanine, and NH_3, which last passes on to form another cyanamide in the chain.

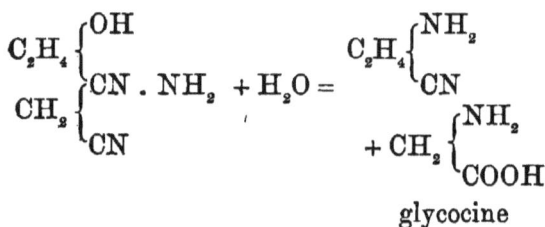

$$C_2H_4 \begin{cases} OH \\ CN \text{ . } NH_2 \\ CN \end{cases} + H_2O = \begin{matrix} C_2H_4 \begin{cases} NH_2 \\ CN \end{cases} \\ + CH_2 \begin{cases} NH_2 \\ COOH \end{cases} \end{matrix}$$
$$\text{glycocine}$$

and

$$C_2H_4 \begin{cases} NH_2 \\ CN \end{cases} + 2H_2O = C_2H_4 \begin{cases} NH_2 \\ COOH \end{cases} + NH_3$$
$$\text{alanine.}$$

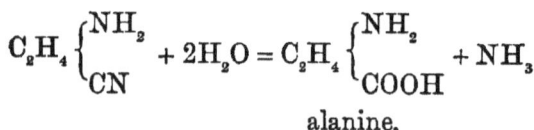

If the larger molecule $\begin{matrix} CH_2 \\ CH_2 \end{matrix} \begin{cases} OH \\ CN \text{ . } NH_2 \\ CN \end{cases}$ is detached from the chain we have then the elements of creatinine $C_4N_3H_7O$, and the transposition of the atoms may be supposed to take place as follows: if the $CN . NH_2$ is detached, and takes the form

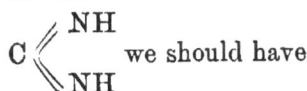

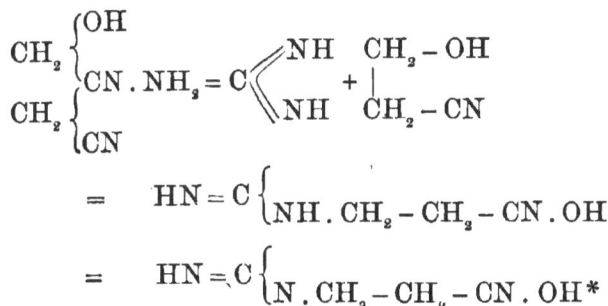

$$C \Big\langle\!\!\!\begin{array}{l} NH \\ NH \end{array}\ \text{we should have}$$

$$\begin{array}{l} CH_2 \\ CH_2 \end{array}\!\!\Big\{\begin{array}{l} OH \\ CN . NH_2 \\ CN \end{array} = C\Big\langle\!\!\!\begin{array}{l} NH \\ NH \end{array} \ +\ \Big|\begin{array}{l} CH_2 - OH \\ CH_2 - CN \end{array}$$

$$=\quad HN = C\Big\{{}_{NH . CH_2 - CH_2 - CN . OH}$$

$$=\quad HN = C\Big\{{}_{N . CH_3 - CH_2 - CN . OH*}$$

But as I have previously pointed out $CN . OH$ represents living, $CO . NH$ dead nitrogen; the formula therefore becomes

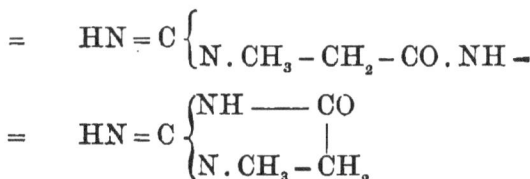

$$=\quad HN = C\Big\{{}_{N . CH_3 - CH_2 - CO . NH -}$$

$$=\quad HN = C\Big\{\begin{array}{l} NH \!-\!\!-\!\!-\!\!- CO \\ N . CH_3 - CH_2 \end{array}$$

The ordinary formula for creatinine †.

The change is perhaps more intelligible if we consider the formation of creatine. If before the separation of the molecule $CN . NH_2$ we hydrate the compound, we have

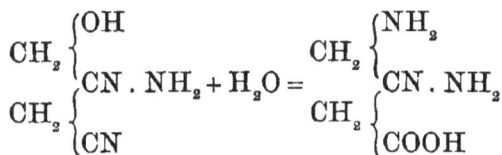

$$\begin{array}{l} CH_2 \\ CH_2 \end{array}\!\!\Big\{\begin{array}{l} OH \\ CN . NH_2 \\ CN \end{array} + H_2O = \begin{array}{l} CH_2 \\ CH_2 \end{array}\!\!\Big\{\begin{array}{l} NH_2 \\ CN . NH_2 \\ COOH \end{array}$$

* We have an example of such a transposition of atoms in the formation of crotonic acid $CH_3 - CH = CH - COOH$ from allyl iodide $CH_2 = CH - CH_2I$ by its conversion into a cyanide and thence into the acid. See Fownes, p. 307.

† Fownes, p. 614.

$$= CN . NH_2 + \begin{vmatrix} CH_2 - NH_2 \\ CH_2 - COOH \end{vmatrix}$$

$$= C \Big\langle \begin{matrix} NH \\ NH \end{matrix} + \begin{vmatrix} CH_2 - NH_2 \\ CH_2 - COOH \end{vmatrix}$$

$$= NH = C \Big\langle \begin{matrix} NH_2 \\ NH . CH_2 - CH_2 - COOH \end{matrix}$$

$$= NH = C \Big\langle \begin{matrix} NH_2 \\ N . CH_3 - CH_2 - COOH \end{matrix}$$

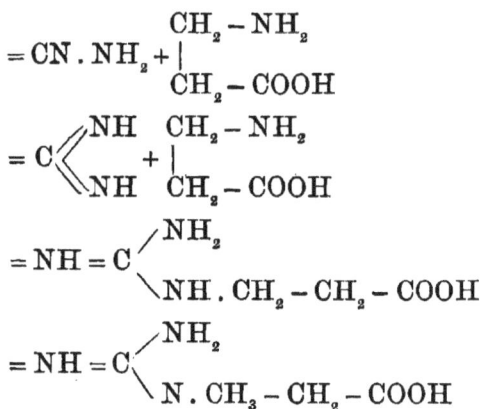

The ordinary formula for creatine.

A third mode in which the molecules may become detached is the following. The ammonia when liberated by hydration of one of the bodies may, instead of combining with the hydroxyl of a cyan-alcohol higher up in the chain to form a cyanamide, combine with the molecule CNOH to form $CNO . NH_4$, which if detached would in the blood as out of the body be converted into urea.

Thus we should have

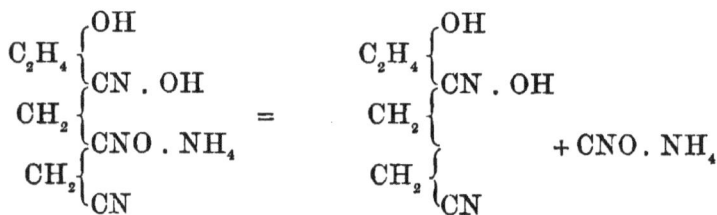

$$C_2H_4 \begin{cases} OH \\ CN . OH \end{cases} \quad C_2H_4 \begin{cases} OH \\ CN . OH \end{cases}$$
$$CH_2 \begin{cases} \\ CNO . NH_4 \end{cases} = CH_2 \begin{cases} \\ \end{cases} + CNO . NH_4$$
$$CH_2 \begin{cases} \\ CN \end{cases} \quad CH_2 \begin{cases} \\ CN \end{cases}$$

and there is formed the next cyan-alcohol in the series
$$CH_2 . CH_2 \begin{cases} OH \\ CN \end{cases} \text{ or } C_2H_4 \begin{cases} OH \\ CN \end{cases}, \text{ and ammonium cyanate or urea*}$$

$$CNO . NH_4 = CO \begin{cases} NH_2 \\ NH_2 \end{cases}$$
$$\text{urea.}$$

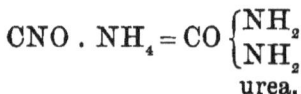

* Fownes' *Chemistry*, p. 96.

In this way we pass from the lower cyan-alcohols to the higher with the formation of urea, the two molecules of CH_2 combining to form C_2H_4. Similarly

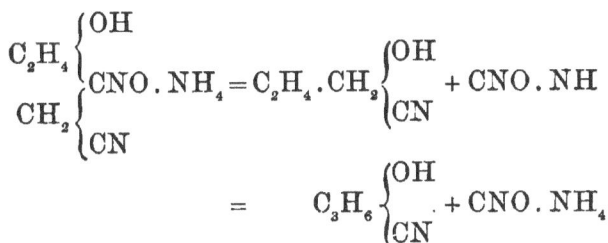

$$C_2H_4 \begin{cases} OH \\ CNO.NH_4 \\ CH_2 \begin{cases} \\ CN \end{cases} \end{cases} = C_2H_4.CH_2 \begin{cases} OH \\ CN \end{cases} + CNO.NH$$

$$= C_3H_6 \begin{cases} OH \\ CN \end{cases} + CNO.NH_4$$

and
$$C_3H_6 \begin{cases} OH \\ CNO.NH_4 \\ C_2H_4 \begin{cases} \\ CN \end{cases} \end{cases} = C_5H_{10} \begin{cases} OH \\ CN \end{cases} + CNO.NH_4$$

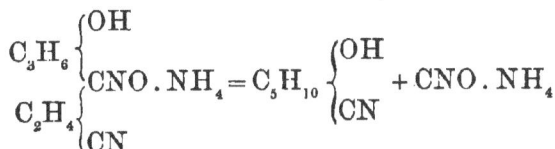

Here we have not only an explanation of the formation of urea in the tissues but the reason why the amido-acids obtained from the tissues possess different properties from those made in the laboratory. It may easily be shown that the last cyan-alcohol $C_5H_{10} \begin{cases} OH \\ CN \end{cases}$, from which leucine may be prepared, will contain six different forms of $C_5H_{10} \begin{cases} OH \\ CN \end{cases}$.

There is another mode in which these molecules may split up, which is interesting from two points of view. It appears to explain the formation of lactic acid and carbonic acid when a muscle contracts or when it dies, and it seems to throw some light on the formation of tyrosine.

If the molecule $\begin{matrix} CH_2 \\ \\ CH_2 \end{matrix} \begin{cases} OH \\ CNOH \\ \\ CN \end{cases}$ were detached, and the

portion CNOH then liberated, the latter would, with H_2O, be decomposed at once into CO_2 and NH_3, the remainder hydrated forming lactic acid

$$\left.\begin{array}{l} CH_2 \left\{\begin{array}{l} OH \\ CNOH \end{array}\right. \\ CH_2 \left\{\begin{array}{l} \\ CN \end{array}\right. \end{array}\right\} + 4H_2O = \begin{array}{l} CH_2 - OH \\ | \\ CH_2 - COOH \\ \text{lactic acid} \end{array} + CO_2 + 2NH_3$$

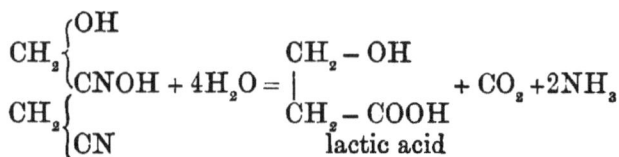

the NH_3 combining with other cyan-alcohols to form cyanamides.

Changes similar to these offer a very simple explanation of the formation of the cyanalcohols in plants, and as a consideration of these changes will help to explain the metabolic changes in the liver, I will briefly discuss them.

In plants CO_2 entering by the leaves combines with H_2O sent up from the roots, and from these starch is said to be formed, a volume of oxygen equal to that of the CO_2 absorbed, being exhaled by the plant. MM. Loew and Bokorny, and Pringsheim have shown that there is a substance in living plasma which has the property of reducing silver salts and for this reason is regarded as an aldehyde. The aldehyde resulting from the combination in the plant of CO_2 and H_2O. Pringsheim gives the following as the reaction; the volume of CO_2 absorbed and of O given out, being the same;

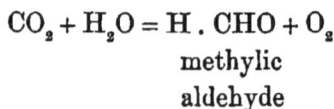

$$CO_2 + H_2O = H . CHO + O_2$$
$$\text{methylic}$$
$$\text{aldehyde}$$

By condensation of the aldehyde glucose is formed

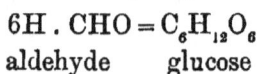

$$6H . CHO = C_6H_{12}O_6$$
$$\text{aldehyde} \qquad \text{glucose}$$

which by dehydration is converted into starch

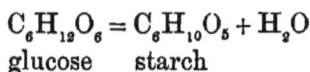

$$C_6H_{12}O_6 = C_6H_{10}O_5 + H_2O$$
$$\text{glucose} \qquad \text{starch}$$

But if instead of undergoing condensation the methylic aldehyde is oxidised it will be converted into formic acid

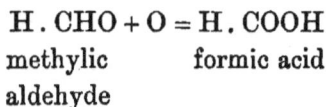

$$H . CHO + O = H . COOH$$
$$\text{methylic} \qquad \text{formic acid}$$
$$\text{aldehyde}$$

This now combines with NH_3 absorbed by the roots forming ammonium formate which, dehydrated, is converted into HCN.

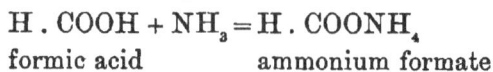

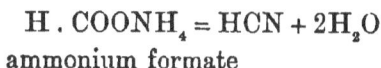

$$H . COOH + NH_3 = H . COONH_4$$

formic acid   ammonium formate

$$H . COONH_4 = HCN + 2H_2O$$

ammonium formate

the HCN now combines with the methylic aldehyde to form the cyan-alcohol $CH_2 \begin{cases} OH \\ CN \end{cases}$ and from condensation of these molecules, with the separation of CNOH, the other cyan-alcohols are formed $C_2H_4 \begin{cases} OH \\ CN \end{cases} C_3H_6 \begin{cases} OH \\ CN \end{cases}$ &c. The CNOH is with H_2O decomposed into CO_2 and NH_3 the latter combining with a fresh molecule of formic acid, to be then transformed as above into HCN.

From these considerations then, and having regard to the mode in which, as I have suggested, urea may be formed, it would appear that when the amido-acids are transformed into cyan-alcohols and pass into the liver condensation takes place with the formation of higher cyan-alcohols, and separation of CNOH and that this change is constantly taking place all through the albuminous chain, the CNOH either combining with NH_3 to form urea or splitting up into CO_2 and NH_3. In this way the glycocine in the alimentary canal would on passing into the liver be transformed into $C_2H_4 \begin{cases} OH \\ CN \end{cases}$ and CNOH, but we have no explanation of the origin of the lowest cyan-alcohol. But if in plants, glucose is a condensation product of methylic aldehyde, it is then no violent assumption that in the liver, the glycogen, after being transformed into glucose, is resolved into its elements of six molecules of methylic aldehyde, and that this, by oxidation, combination with NH_3 and dehydration, is transformed as in plants into the lowest cyan-alcohol of the series, viz. $CH_2 \begin{cases} OH \\ CN \end{cases}$. This view is supported by the experiments of Nägeli* which

* *Sitzungsb. d. Bayr. Acad. d. Wiss.* 1879.

showed that albumen can be formed by fungi whose nutriment consists exclusively of sugar mixed with ammonia.

Now tyrosine when fused with potash is resolved into acetic acid, ammonia and para-oxy-benzoic acid*. As the other derivatives of albumen can be obtained from the various aldehydes, should not this substance also be derived from the corresponding aldehyde? By combining oxy-benzoic aldehyde or salicylal† with HCN we should have

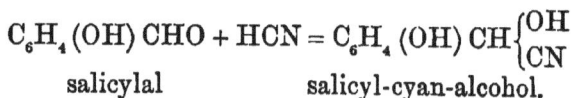

$$C_6H_4(OH)CHO + HCN = C_6H_4(OH)CH\begin{cases}OH\\CN\end{cases}$$

salicylal   salicyl-cyan-alcohol.

Combining this with methene cyan alcohol we have

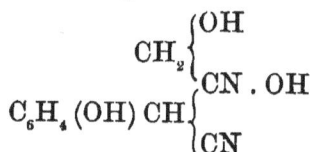

$$C_6H_4(OH)CH\begin{cases}CH_2\begin{cases}OH\\CN.OH\end{cases}\\CN\end{cases}$$

which hydrated would form

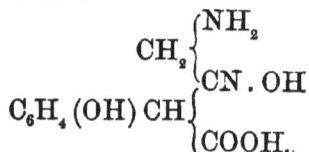

$$C_6H_4(OH)CH\begin{cases}CH_2\begin{cases}NH_2\\CN.OH\end{cases}\\COOH.\end{cases}$$

If now the molecule CNOH were detached, forming CO_2 and NH_3, the substance would then be converted into tyrosine

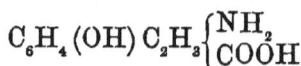

$$C_6H_4(OH)C_2H_3\begin{cases}NH_2\\COOH\end{cases}$$

so that by the condensation of salicyl-cyan-alcohol and methene cyan-alcohol, with elimination of CNOH, we arrive at the constitution of tyrosine. Here then we have the molecule of CNOH which was wanting in our formula for albumen. If therefore in

---

* Fownes, p. 539.

† Salicylal is ortho-oxy-benzoic aldehyde—but benzene compounds sometimes undergo isomeric change when distilled with potassic cyanide. Compounds from which resorcin, a meta-derivative of benzene, is obtained when distilled with potassic cyanide can furnish para-derivatives of benzene, viz. terephthalic acid. See Miller's *Org. Chem.*, 1880, Part III. p. 526.

the formula on page 21 we substitute for $C_2H_3 \begin{cases} OH \\ CN \end{cases}$

$$CH_2 \begin{cases} OH \\ CNOH \end{cases}$$
$$CH \begin{cases} \\ CN \end{cases}$$

we have as our formula for albumen

$C_5H_{10} \begin{cases} OH \\ CNOH \end{cases}$
$C_5H_{10} \begin{cases} CNOH \end{cases}$
$C_5H_{10} \begin{cases} CNOH \end{cases}$
$C_4H_8 \begin{cases} CNOH \end{cases}$
$C_4H_8 \begin{cases} CNOH \end{cases}$
$C_4H_8 \begin{cases} CNOH \end{cases}$
$C_3H_6 \begin{cases} CNOH \end{cases}$
$C_3H_6 \begin{cases} CNOH \end{cases}$
$C_3H_6 \begin{cases} CNOH \end{cases}$
$C_2H_4 \begin{cases} CNOH \end{cases}$
$C_2H_4 \begin{cases} CNOH \end{cases}$
$C_2H_4 \begin{cases} CNOH \end{cases}$
$CH_2 \begin{cases} CNOH \end{cases}$
$CH_2 \begin{cases} CNOH \end{cases}$
$CH_2 \begin{cases} CNOH \end{cases}$
$CH_2 \begin{cases} CNOH \end{cases}$
$CH_2 \begin{cases} CNOH \end{cases}$
$CH \begin{cases} CNOH \\ CN \end{cases}$

a substance whose composition is $C_{72}H_{118}N_{18}O_{22}S$, differing from Lieberkühn's formula $C_{72}H_{112}N_{18}O_{22}S$ only by six atoms of hydrogen.

I have thus endeavoured to show that albumen is a compound of cyan-alcohols united to a benzene nucleus, these being derived from the various aldehydes, glycols and ketones, or that they may be formed in the living body by the dehydration of the amido-acids; that from a body so constituted all the different substances may be obtained which have been extracted from albumenoid tissues; that lactic acid is obtained in two ways, either from $C_2H_4 \begin{cases} OH \\ CN \end{cases}$, or from changes and condensation in $CH_2 \begin{cases} OH \\ CN \end{cases}$, with the simultaneous development of CO_2, a result which is brought about when a muscle contracts or when it dies; and that urea may be obtained from one series of cyan-alcohols with the production of a cyan-alcohol higher in the series.

Such a compound of cyan-alcohols therefore presenting so much resemblance in its properties to albumen, cannot I think differ very widely (though perhaps not absolutely correct) from the molecular constitution of albumen.

Taking this view then of the constitution of albumen the following may be given as a summary of the nutritive changes.

The amido-acids glycocine, leucine, tyrosine, &c. in passing from the alimentary canal to the liver are dehydrated, forming a series of cyanhydrins or cyan-alcohols attached to a benzene nucleus, and then pass into the circulation. In the tissues these cyan-alcohols, partly by condensation, partly by hydration and oxidation, give rise to the various effete products which are eliminated from the system chiefly in the form of carbonic acid and urea.

There is still one other point that I do not wish to leave unnoticed. According to Tiemann, the cyanamides $R - C \begin{cases} NH_2 \\ \longleftarrow H \\ CN \end{cases}$

are very unstable bodies and with the elimination of NH_3 very easily condense into Imido-nitriles*.

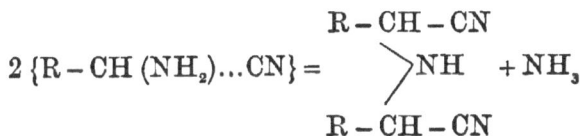

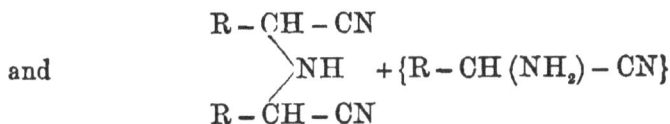

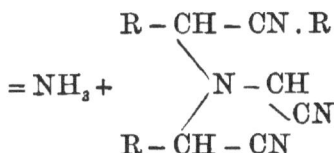

$$2\{R-CH(NH_2)\ldots CN\} = \begin{array}{c} R-CH-CN \\ \diagup NH \\ R-CH-CN \end{array} + NH_3$$

and

$$\begin{array}{c} R-CH-CN \\ \diagup NH \\ R-CH-CN \end{array} + \{R-CH(NH_2)-CN\}$$

$$= NH_3 + \begin{array}{c} R-CH-CN.R \\ \diagdown N-CH \\ \diagup \diagdown CN \\ R-CH-CN \end{array}$$

If then the force holding the cyan-alcohols composing living proteid together were suddenly withdrawn, changes would quickly take place in these unstable bodies, there would be the formation of some acid, and the different cyanamides, which latter would undergo the condensation above described; the liberated NH_3 combining with other cyan-alcohols to form other cyanamides and further condensation taking place. Does this not offer some clue to the phenomena of rigor mortis and the coagulation of the blood?

Again, by combining two molecules of $CH_2\begin{Bmatrix} NH_2 \\ CN \end{Bmatrix}$ we should have by Tiemann's formula

$$2CH_2\begin{Bmatrix} NH_2 \\ CN \end{Bmatrix} \quad \begin{array}{c} CH_2-CN \\ \diagup NH \\ CH_2-CN \end{array} + NH_3$$

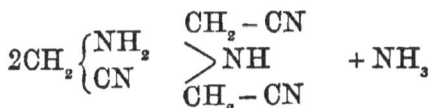

which when hydrated with weak acids would give

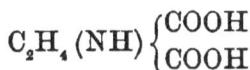

$$C_2H_4(NH)\begin{Bmatrix} COOH \\ COOH \end{Bmatrix}$$

---

* Berichte, xiv. s. 1958.

a body having the same composition as aspartic acid, but differing
in that it is an *imido* instead of an *amido* body.   Similarly

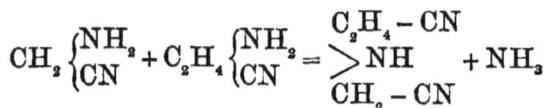

$$CH_2 \begin{cases} NH_2 \\ CN \end{cases} + C_2H_4 \begin{cases} NH_2 \\ CN \end{cases} = \begin{matrix} C_2H_4 - CN \\ > NH \\ CH_2 - CN \end{matrix} + NH_3$$

which hydrated would give $C_3H_6(NH) \begin{cases} COOH \\ COOH \end{cases}$ a body having the

same composition as glutamic acid, differing only in structure.
But with strong HCl and high temperature these 'imido-nitriles'
seem always to give aldehyde, HCN and an amido acid.   It is not
improbable therefore that under certain conditions we may have

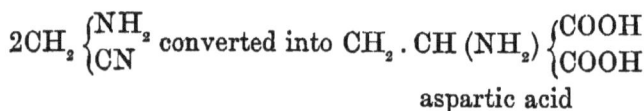

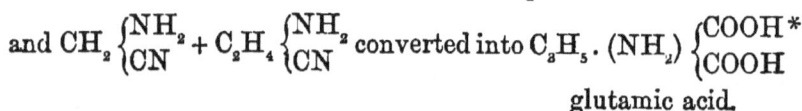

$$2CH_2 \begin{cases} NH_2 \\ CN \end{cases} \text{ converted into } CH_2 . CH(NH_2) \begin{cases} COOH \\ COOH \end{cases}$$
aspartic acid

and $CH_2 \begin{cases} NH_2 \\ CN \end{cases} + C_2H_4 \begin{cases} NH_2 \\ CN \end{cases}$ converted into $C_3H_5 . (NH_2) \begin{cases} COOH* \\ COOH \end{cases}$
glutamic acid.

In discussing the pathology of rheumatism, gout and diabetes,
we shall have to deal chiefly with the lowest cyan-alcohol of the
series, viz. $CH_2 \begin{cases} OH \\ CN \end{cases}$, and the substances derived from it.   What
part these play in regard to these disorders I will endeavour to
show in the next lectures.

* We have such a change of an imido into an amido body in the action
of boiling hydrochloric acid on hydrazo-benzoic acid. See Fownes, p. 526.

LECTURE II.

WHAT in Rheumatism is the starting point of the morbid process? Is there such a thing as the rheumatic poison, if so what is it and what is the nature of the change which takes place as the result of its action? Let me in the first place enumerate the more salient features of the disease as gleaned from clinical experience. An individual after feeling out of sorts for two or three days, perhaps after a chill, wakes up in the morning with one of the joints stiff and tender; soon afterwards it swells, and motion causes pain. Other joints now become affected; the knees, wrists, elbows, ankles, and the smaller joints of the hands and feet. There are pain, tenderness, increased heat, swelling, and redness of the skin. The tendency of these to shift from joint to joint is a remarkable feature of the disease. The slightest motion becomes insupportable, even vibration of the room or of the bed causes suffering. There is more or less febrile movement, the temperature ranging from 102^{0} to 104^{0} or higher, generally more or less sweating, the odour of which is characteristic. The tongue is coated; thirst is generally a marked symptom and the urine is turbid with lithates and has a strong acid reaction.

The more acute and severe the case the more profuse and acid is the perspiration, and the larger the amount of lactic acid which it contains. This last feature has led to the supposition that in acute rheumatism there is excessive formation of this acid which, circulating through the system, gives rise to the symptoms which accompany the disorder. It is a theory which has got a very firm hold of the minds of many physicians, and is probably the one

3—2

most generally adopted. There is so much to be said in its favour that until recently I unhesitatingly accepted it. I hope to show now that, though lactic acid is formed in excess in the disease, and its presence in the blood modifies the symptoms in some degree, it is not the chief agent in producing them. Other factors are at work giving rise to the characteristic feature of rheumatism, namely the affection of the joints, as well as to some of the other symptoms which I have enumerated; and the excessive formation of this acid must be regarded more as one of the symptoms, than as the cause which gives rise to them.

The lactic acid theory however, has furnished me with my true starting point. In attempting five or six years ago to arrive at some satisfactory explanation of the phenomena of this disorder, the line of argument which suggested itself to my mind was much as follows.

1.   Lactic acid is the *materies morbi* in rheumatism,

2.   According to the views of some authorities the natural source of lactic acid in the system is glucose, which is first transformed into lactic acid, and then further oxidised into carbonic acid and water.

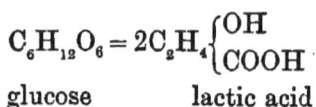

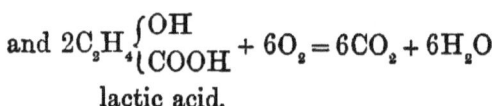

$$C_6H_{12}O_6 = 2C_2H_4\begin{cases}OH \\ COOH\end{cases}$$

glucose   lactic acid

$$\text{and } 2C_2H_4\begin{cases}OH \\ COOH\end{cases} + 6O_2 = 6CO_2 + 6H_2O$$

lactic acid.

3.   By the so-called 'diabetic puncture' the metabolism of glucose in the system can be arrested—sugar appears in the urine and we have the phenomena of diabetes.

Changes therefore in the medulla oblongata entirely arrest the metabolism of glucose and if so, if this is true, would it be possible by modifying these changes in the medulla partially to arrest the transformation of glucose into CO_2 and H_2O at the inter mediate stage of lactic acid, andso develope rheumatism?

That is to say, have neurotic conditions, have changes in the medulla oblongata anything whatever to do with the symptoms which are associated with rheumatism?

These were the hypotheses with which I started; but I have tried to show in my previous lecture, that lactic acid in the tissues is formed, not from glucose but from the nitrogenous molecules of the tissues, namely the cyan-alcohols $CH_3 . CH \begin{cases} OH \\ CN \end{cases}$ and $CH_2 CH_2 \begin{cases} OH \\ CN. \end{cases}$

My second hypothesis therefore falls to the ground, and with it the third to a very large extent. Certain points however presented themselves to me during the consideration of the subject, which seem to connect the disorder with changes in the nervous system, and to furnish a theory as to the causation of rheumatism, by which an explanation may be offered not only of the way in which the symptoms are produced, but also of the action of some of the remedies which have been successfully employed in the treatment of the disease. To these points let me now specially invite your attention.

As a first step in demonstrating the connection of rheumatism with changes in the nervous system, let me refer to some nerve lesions which are associated with changes in the nutrition of the joints. In respect to this I find that more than fifty years ago the late Dr J. K. Mitchell* advanced a theory of the pathogenesis of rheumatism which connected it with affections of the spinal cord. My own view which localises the controlling action of the nervous system in the disorder, chiefly in the medulla oblongata, was formed quite independently, it was only in fact quite recently that my attention was directed to Dr Mitchell's views by finding a reference to them on reading a most admirable work, abounding with interesting facts, by his son Dr Weir Mitchell on "Injuries of nerves." I here discover however some most important links in the chain of evidence supporting my theory, which I will place before you; and after referring to other cases of a similar kind I shall then attempt to point out the way in which the nervous

* *The American Journal of Medical Sciences*, 1831, Vol. VIII. p. 55.

centre is affected both in gout and rheumatism—how in fact the
rheumatic poison is developed and how it acts.

In a work on *"Gunshot Wounds and other Injuries of Nerves"*
by Drs Weir Mitchell, Moorhouse and Keen of the United States
Army * there is the following statement:—

"Alterations in the nutrition of joints. Again we call attention
to a peculiarity of nerve injuries which we believe to have been
overlooked.

"Like the altered nutrition of the skin, the symptom which we
are at present considering occurs at any time after the first few days.
It consists essentially in a painful swelling of the joints, which
may attack any or all of the articulations of a member. It is distinct from the early swelling due to the inflammation about the
wound itself, although it may be masked by it for a time; nor is it
merely a part of the general œdema which is a common consequence of wounds. It is more than these—more important, more
persistent. Once fully established, it keeps the joints stiff and sore
for weeks or months. When the acute stage has departed, the
tissues about the articulations become hard, and partial anchylosis
results, so that in many cases the only final cause of loss of motion
is due to this state of the joints. Of all the agencies which im-
pede movement, it is the most difficult to relieve.

"Were we asked to state in what essential respect these lesions
differ from subacute rheumatic disease of the same parts, we
should certainly be at some loss to discern a difference.

"The subject suggests certain interesting reflections. We have
ourselves seen cases of spinal injury, in which rheumatic symptoms seemed to have been among the consequences; and four such
instances of striking character are to be found recorded in a paper
by the late Prof. J. K. Mitchell, *Am. Jour. of the Med. Sci.* vol.
viii. p. 55 †. Upon the hints which were thus furnished, Dr

* * *

* *Gunshot Wounds and other Injuries of Nerves*, by Drs Weir Mitchell,
Moorhouse and Keen. Philadelphia, 1864, pp. 83—85.

† The treatment he adopted was to apply from eight to sixteen cups,
abstracting about as many ounces of blood, to the cervical or lumbar spine

Mitchell was induced to consider rheumatism as of spinal origin. His treatment, founded on this view, was most successful, and is still used and recognised in this country. Modern pathologists have traced the causation of rheumatism to a strictly chemical source; but no one can avoid seeing that there may be a cause beyond this, even though the chemical conditions be still considered as essential links in the chain. Thus, after all, the true origin may be spinal, or, at all events, the indisputable fact that there are rheumatisms depending for existence on neural changes, may aid us hereafter to discriminate varieties of type among the forms of rheumatic disease. It were easy to dwell upon this subject, but enough has been said to show that subacute inflammation of joints may be brought about by nerve lesions, and to direct medical thought anew in a direction which seems favourable to its true and rational progress."

In a subsequent work—the work to which I have already referred—"*On Injuries of Nerves*"* Dr S. Weir Mitchell writes: "Of all the various forms of mischief wrought by nerve wounds, the most intractable and disabling are the curious inflammatory states of joints to which we were the first to call attention."......

"In a certain number of nerve wounds, notably most often in those of the upper extremities, one or more of the joints in the wounded limb become swollen. The nature of the injury does not seem to influence the case, as I have seen it follow dislocations, ball wounds, and contusions of nerves, while in an interesting case of Dr Packard's, it was one of the consequences of compression of the sciatic nerve by a tumour. More lately, in the service of my friend Dr J. A. Brinton, at the Philadelphia Hospital, I saw a man who had extensive joint lesions, owing to the brachial nerves having suffered during the dislocation, or upon the subse-

according as the upper or lower extremities were affected and follow the cupping, if this did not afford relief, by the application of blisters to the spine. This paper appears to be of so much interest and importance, that I have added it as an Appendix to these lectures; see page 107.

* London and Philadelphia, 1872, pp. 168—172.

quent reduction of the humerus, so that I suspect these troubles are
more common than has been supposed*. In one case the joints of
the fingers became swollen and tender on the third day after ball
wound of the brachial plexus, but usually the swelling appears
much later, and, like the glossy skin, is frequently the offspring of
secondary neuritis. Often masked at first by the general inflammation of the limb, or concealed by the œdema so common after
nerve wounds, it is more persistent than these, and, as they fade,
begins to assume importance. We may then have one articulation
—and if only one, a large one—involved, or perhaps all the joints
of a finger, or every joint in the hand, or of the entire limb may
suffer. The swelling is never very great, the redness usually
slight, and the tenderness on touch or motion exquisite. This condition of things remains with little change during weeks or months,
and then slowly declines, leaving the joints stiff, enlarged, and
somewhat sensitive, especially as to movement. A small proportion of such cases find ready relief, but in many of them the
resultant anchylosis proves utterly unconquerable, so that it is
vain to break up the adhesions under ether, or to try to restore
mobility by manipulation or splints. All alike fail, and serve only
to add to the essential tortures of the accompanying neuralgia and
hyperæsthetic states of skin. Since writing my last paper † I have
met with some of the former patients who suffered with these
troubles, but in no case originally very severe was there any great
gain,—indeed in most of them the joints had become every year
more stiff and useless.

"It is then quite clear that injuries of the spine, diseases of this
organ, and of the brain, and wounds, or any form of lesions of
nerves, are capable of developing in the joints inflammatory conditions, usually subacute, and which so precisely resemble rheumatic arthritis in their symptoms and results, that no clinical skill
can discriminate between the two. In this state it were well to
leave the subject. The chemical theories have crumbled, and, in

---

* I have since met with similar cases.
† Reports of the Sanitary Commission.

the growing tendency to believe that rheumatism may have more forms than one, it may not be amiss to recall the facts to which we have contributed, and which are well illustrated by the following case. Other and more severe examples will be found in the cases appended to the later chapters of this work.

"*Case* 30.—*Gunshot wound of the right brachial plexus; causalgia; tremor; arthritic lesions; nail-changes; acid sweats; hyperæsthesia; little loss of motion from paralysis; great gain under treatment.* B. D. L., aged forty-three, a farmer from Maine. Enlisted July, 1862. He was healthy to the date of his wound, received July 2d, 1863, at Gettysburg. While kneeling and aiming he was shot in the right side of the neck. He felt pain in the wound, but none down the arm. He spun around, feeling stunned, and fell on his back, not unconscious. In five minutes he arose and walked to the rear, where the wound was dressed with cold water, no splint being employed either then or later. At first, all motion was lost. In an hour he could move his finger and abduct the arm, but not flex it. He thinks sensation was perfect, except as to the ulnar distribution. Within an hour he had severe earache, and pain in the shoulder, arm, and forearm. During the second week he began to have burning pain in the hand. At this time, which probably marked the onset of neuritis, the shoulder-joint grew stiff, then the elbow, and lastly all of the fingers. This condition was excessively painful, and remained unchanged. The tremor which is constant in the upper arm muscles began the day of the wound, and had not ceased on his admission to our wards.

"*Site of wound.*—On admission, October, 1863, it was noted that the ball had entered the right side of the neck, in front, three inches above the clavicle, in the outer edge of the trapezius. The missile passed downward and outward, and struck the anterior edge of the supra-spinal fossa of the scapula, five inches external to the spine of the first dorsal vertebra. Both wounds sloughed, leaving scars one and a half inches in diameter. The patient is well and florid. The shoulder is motionless from stiffness. The lower joints are alike stiff, swollen, red, and painful; the arm, semi-prone and flexed, is carried across the chest, supported by the sound hand. He has slight motion throughout, but the effort causes fibrillar-tremor and exquisite pain.

"*Sensation.*—The sense of touch is everywhere good, save that there is slight numbness of the back of the hand and forearm. Some causalgia is felt in the palm, but no other pain, except on movement.

"*Nutrition.*—The palm is thin and red or purplish, and on it the patient uses water, now and then, as a dressing; there is no atrophy; the wound is healed, but tender, as are also the upper nerve tracks. Muscular hyperæsthesia of the deltoid and triceps is present. The nails are remarkably curved; the hair is scanty; the sweat ill smelling and acid. The shoulder muscles alone have lost electro-muscular contractility (induction current). Under ether, the joints when moved are found to be free from well-marked organic adhesions.

"Passive motion and electricity caused speedy gain in movement, and in February, 1864, he was able to move all the joints with diminished pain. The muscles were, at this time, sensitive to induced currents, and the numbness and causalgia had nearly disappeared. He was allowed a furlough, at the expiration of which he deserted.

"This case is valuable as an example of arthritic changes, extreme in character, with very little sensory or motor paralysis, and seemingly aided by treatment."

Again Sir William Gull* has recorded the two following cases, and his remarks upon them are so suggestive, that I quote them in full.

"*Case* 27. *Acute rheumatic* (?) *affection of the larger joints. Paraplegia of lower extremities. Slough over sacrum. Recovery.* Anne E—, æt. 39 was admitted into Guy's Hospital, March 31st, 1857, under the care of my colleagues, Dr Hughes and Dr Wilks (to whom I am indebted for placing the case at my disposal).

* Guy's *Hosp. Reports*, 3rd Series, Vol. iv. 1858, Cases 27 and 28.

Both hands were swollen, stiff and painful with an erythematous blush over the back of the right, and on the second joint of the thumb of the left. The legs were so far paralysed that she could only very slowly and feebly move them. The muscles were greatly wasted and flabby, but had not lost their excito-contractility by galvanism. Sphincters weak. No swelling of the knees or ankles at this time. Sensation nearly normal, but at times both legs felt numb, and were drawn up involuntarily. Urine acid, high coloured and scanty. Tongue covered with a cream-like fur; skin hot, perspiration profuse, with acid smell. Pulse 120; systolic murmur over ventricle. On examining the spine, the lower third of the sacrum was found to be bent forward, the result of a fall eleven years before; and near the sacral notch, on the right side, was the cicatrix of a wound which formed at that time. Except this, there was nothing abnormal, nor any pain or tenderness on pressure. The history she gave of her case was, that being a widow, she was necessitated to work laboriously at a mangle. She had for two years, when much exerting herself, felt pain in the back between the shoulders, and a sense of constriction and coldness round the chest. Ten days before coming into the hospital, she was seized with pain in the left leg, and had spasmodic contraction of the muscles, with an increase of the pain and constriction round the chest. She had still power to extend the leg but could not walk. The day following, the hand, knees and ankles were swollen and painful. With these symptoms there was febrile heat and diarrhœa. The sphincter ani was so weak that the feces ran from her involuntarily. On the third day a slough formed over the sacrum. No important change occurred in her symptoms after her admission. There was great muscular emaciation generally. Involuntary twitchings of the muscles of the arms and legs. Aching, gnawing sensations in both calves. Touching the feet gave rise to formication, and very lively excito-motor movements. For ten days the hands remained red, painful, stiff and swollen. She complained much of heat and profuse perspirations, which returned several times

in the twenty-four hours. On the 8th of April the urine was
ammoniacal, and contained mucus. The hands were still swollen
and erythematous; face flushed; pulse 100, full, as in rheumatism; acid smell of perspiration; respiration 28; movements
thoracic and abdominal; abdomen soft; pupils large, nights sleep-
less. Ordered a grain of opium every six hours, with six ounces
of wine daily, and a chop. On April 18th the good effects of
the opium and support were very apparent. The patient had
passed good nights, and was tranquil in the day. Perspiration
lessened. Urine retained in the bladder for thirty-six hours was
at length passed voluntarily, it was acid and without mucus.
Tongue pale and moist. The slough on the back had deepened.
The pupil still continued large. Occasional contraction of the
muscles of the legs. No permanent rigidity. Hands remained
swollen and stiff, but less red. She was unable to move the
shoulders freely. On April 22nd, the hands had recovered their
normal appearance, and had lost their stiffness. The legs could
be moved more freely. The sense of constriction round the chest
was gone; pulse 96; skin cool and dry; appetite good; urine
normal, but she could not empty the bladder oftener than once in
twenty-four hours. From this date she slowly recovered. The
opium was continued throughout her convalescence. At the
beginning of June the muscles of the lower extremities were
galvanized regularly. By the end of the month she was able to
stand without help. Her improvement was uninterrupted, and
in September, she left the hospital quite well.

"*Remarks.* It is a matter of great clinical interest that lesions
of the cord are occasionally attended with an affection of the
joints not to be readily distinguished from that which occurs in
acute rheumatism. When this happens there may be difficulty in
determining the pathology of a case. It may, indeed, be impossible to say whether the symptoms at a certain stage are due to
disease of the cord, or to a rheumatic state of the blood. In
such instances we have a proof of the near relation of humoralism
and solidism; for one observer may maintain that the local lesions

have a common origin in the altered state of the blcod, whilst another may with equal confidence assert their dependence upon a primary disturbance of the nervous centres. The case here recorded is an example of these difficulties. Fatigue from mechanical labour, acting especially on the lumbar and dorsal portions of the spine in a delicate and anxious subject, appears to have injured the nutriment of the cord. For two years, when much exerting herself, the patient felt pain between the shoulders, and a sense of constriction and coldness round the chest. Paraplegia then suddenly came on, followed by redness, pain, and swelling of the larger joints, as in rheumatism. Together with these symptoms, there were others indicating a rheumatic condition—white furred tongue; flushed face; hot skin; profuse perspirations, having an acid smell; systolic murmur over left ventricle, &c. Was there here a rheumatic state of the blood induced by the spinal lesion; or was the nervous derangement the result of a rheumatic state? Notwithstanding the labours of morbid anatomists and chemical pathologists, we are not at present in possession of any certain knowledge of what constitutes the rheumatic condition. My colleague Dr Addison, from his clinical experience, has long drawn attention to the close connection between spinal lesions and true rheumatism, but has never developed the idea beyond expressing a suspicion of their relation. At the time this case was under care the treatment was a subject of much observation. The result was very satisfactory. Whatever might have been the state of the cord, it was clearly induced by fatigue, and was soon followed by sloughing of the integuments. It would not, therefore, admit of depletory measures, but on the contrary, required a nutritious diet, and wine. Opium was prescribed apparently with great advantage; it allayed nervous irritability, and gave the patient sleep.

"The following case is also illustrative of the relation between spinal injury and rheumatic symptoms. The same plan of treatment as above was equally successful. The therapeutical view of this subject is certainly not without the greatest interest. No

doubt the texture of the cord has but feeble reparative powers,
notwithstanding it has been shown by experimenters on animals,
that occasionally, after a transverse section, the parts unite, and
the functions are re-established.

"*Case* 28. *Concussion of the spine; partial paraplegia; redness
and swelling of the wrists and ankles as in acute rheumatism.
Recovery.* N. T.— æt. 38, on the 22nd January, 1855, inad-
vertently stepped backwards into a hole, a few feet deep, and
received a concussion of the spine. After a few days he became
partially paraplegic, with weak sphincters; and at the same time
there came on a diffused redness and swelling of the ankles and
wrists. The swelling was not from effusion into the joints, but
from œdema of the surrounding tissues. The joints were very
painful. The redness and swelling were variable in degree.
When most marked they presented the usual appearances of
rheumatism, or rather of gout, for the erythema was brighter,
and the œdema more distinct than in rheumatism. The hands
equally affected with the ankles, though there was no obvious
want of muscular power, nor any affection of sensation in the
upper extremities. Tongue clear. Pulse 120. No acid perspirations. Urine high coloured, free from deposits; of normal quantity. The nerves of the surface generally were hyperæsthetic to a
slight touch, but deep pressure gave less inconvenience. The
treatment consisted of good nourishment, wine and brandy freely
administered, and opium to allay pain and overcome sleeplessness.
The pulse gradually acquired more power, and sank to 80. The
affection of the joints continued in varying degree through March,
April, May and June. From the beginning of April there was
an improvement in the power over the legs. The same treatment
was continued throughout, without the use of mercurials, local
depletion, or counter irritation. In June, the patient was able to
walk without assistance. During sleep, the hands and feet, wrists
and ankles often became erythematous and swollen. There was
occasional formication in the lower extremities. Sleeplessness,
from the beginning of the case, and throughout was a troublesome

symptom. In July, the patient was able to leave the hospital, and to resume to some extent his duties as medical practitioner. He was under the care of my colleague Mr Cock."

Again. According to Professor Charcot* "nutritive disorders consecutive on lesions of the nervous centres not infrequently take up their seat in the articulations." He establishes two varieties. The first comprising such cases as those I have already referred to,—and including here also the *arthropathy of hemiplegic patients*, first described in 1846 by Scott Alison †, afterwards by Brown-Séquard ‡; these "arthropathies are limited to the paralysed limbs, and mostly occupy the upper extremities. They supervene, especially after circumscribed cerebral ramollissement, and more rarely as a consequence of intra-encephalic hæmorrhage.

"They usually form fifteen days or a month after the attack of apoplexy, that is to say at the moment when the *tardy contracture* that lays hold on the paralysed members appears, but they may also shew themselves at a later epoch. The tumifaction, redness, and pain of the joints are sometimes marked enough to recall the corresponding phenomena of acute articular rheumatism. The tendinous sheaths are, indeed, often affected at the same time as the articulations.

"It is needless to insist upon the interest which pertains to these arthropathies as regards diagnosis,—articular rheumatism, whether acute or subacute, being an affection often connected with certain forms of cerebral softening, and one which, indeed, shows itself also, occasionally, after traumatic causes capable of determining shock in the nervous centres. On the other hand, many affections of the spinal cord are erroneously attributed to a rheumatic diathesis in consequence of the coexistence of these

---

* *Lectures on Diseases of the Nervous System.* Translated by Dr Sigerson. New Sydenham Society, 1877, p. 92.

† "Arthritis occurring in the course of Paralysis". *The Lancet*, Vol. i. p. 278, 1846.

‡ *Ibid.* July 13, 1861. See Appendix also, p. 117, where Charcot's criticism on the views of Alison and Brown-Séquard will be found.

articular symptoms. The clinical characters which render it easy to recognise arthropathies correlated with lesions of the nervous centres, and which allow them to be distinguished from cases of rheumatic arthritis are chiefly these:

"1st. Their limitation to the joints of the paralysed members.

"2nd. The generally determinate epoch in which, in cases of sudden hemiplegia, they make their appearance on the morbid scene.

"3rd. The coexistence of other trophic troubles of the same order, such as eschars of rapid formation; and (when the spinal cord is involved) acute muscular atrophy of the paralysed members cystitis, nephritis, &c.

"The type of the second group is to be found in progressive locomotor ataxia...

"Ataxic arthropathy usually occupies the knees, shoulders, and elbows; it may also take up its seat in the hip-joint."

Professor Charcot then goes on to describe the physical and clinical characters of this form of arthropathy, and after pointing out the absence of all traumatic or diathetic cause of rheumatism or of gout he proceeds as follows:—

"It is not very rare to find the spinal grey matter affected in locomotor ataxia; but the lesion is then generally found in the posterior cornua. Now, it was quite different in two cases of locomotor ataxia, complicated with arthropathy, in which a careful examination of the cord has been made; the anterior cornua were, in both cases, remarkably wasted and deformed, and a certain number of the great nerve-cells, those of the external group especially, had decreased in size, or even disappeared altogether without leaving any vestiges. The alteration, besides, showed itself exclusively in the anterior cornua, corresponding to the side on which the articular lesion was situated. It affected the cervical region, in the first case, where the arthropathy occupied the shoulder; it was observed, a little above the lumbar region, in the second case which presented an example of arthopathy of the knee. Above and below these points, the grey matter of the anterior cornua appeared to be exempt from alteration.

"From what precedes, I hope to have made it appear at least highly probable that the inflammatory process, first developed in the posterior columns, by gradually extending to certain regions of the anterior cornua of the grey matter was able to occasion the development of the articular affection in our two patients. If the results obtained in these two cases are confirmed by new observations, we should be naturally led to admit that arthritic affections connected with myelitis, and those observed to follow on cerebral softening, are likewise due to the invasion of the same regions of the grey matter of the spinal cord. In cases of brain-softening, the descending sclerosis of one of the lateral columns of the spinal cord might be considered as the starting-point of the diffusion of inflammatory work."

From these illustrations, gathered from various authorities, it is, I think, very clearly shown, that lesions in the spinal cord, or along the course of the motor nerves, may, in many cases, give rise to changes in the joints, which present no difference from those observed in subacute rheumatism; in others, as in Charcot's disease, there is rapid disintegration of the bony tissue associated with the lesion; decay without repair, as it has been aptly termed by Morrant Baker. But, in all the cases, we have changes taking place in the condition of the joints, and these changes produced by causes acting on the nerves connected with the part, either along the course of these nerves, or at their origin in the central nervous system.

Now, is there in the phenomena of rheumatism anything suggestive of an irritating cause acting upon portions of the central nervous system? If so, what is it, and how is it produced?

Let me call your attention to the changes produced in the system when under the normal state of things, the surface of the body is exposed to the action of cold or heat. The effect, as is well known, is very different as regards cold-blooded and warm-blooded animals. External cold diminishes and heat increases the metabolic activity of the cold-blooded animal, acting like a mixture of dead substances in a chemical retort. But in a warm-blooded animal, within certain limits, cold increases and heat diminishes the bodily metabolism, as shown by the increased or diminished consumption of oxygen and production of carbonic acid, as the temperature falls or rises. There is obviously a mechanism of some kind counteracting, and indeed overcoming, those more direct effects which alone obtain in cold-blooded animals. But under the influence of urari, which paralyses the end-organs of the muscular nerves, a warm-blooded animal behaves as regards its bodily metabolism under the influence of external cold and heat, like a cold-blooded animal. A similar result is obtained by division of the medulla oblongata. The temperature of an animal so operated upon falls and then exposure to heat augments and exposure to cold diminishes its metabolic activity. "We can best explain these results by supposing that under normal conditions the muscles which, as we have seen, contribute so largely to the total heat of the body are placed, by means of their motor nerves and the central nervous system, in some special connexion with the skin, so that a lowering of the temperature of the skin leads to an increase, while a heightening of the temperature leads to a decrease, of the muscular metabolism. Further, though the matter has not been fully worked out, the centre of this thermo-taxic reflex mechanism appears to be placed above the medulla oblongata, possibly in the region of the pons Varolii. When urari is given the reflex chain is broken at its muscular end, when the spinal cord is divided the break is nearer the centre*."

How does this bear upon rheumatism? An individual is exposed to a cold draught or gets wet; the surface is chilled; the cutaneous vascular areas are constricted; by reflex action through the vaso-motor system, the splanchnic vascular areas and the vessels of the muscular areas are dilated; the vaso-motor nerves distributed to these parts are paralysed, so to speak; with this

* Foster's *Physiology*, 4th ed. p. 467.

paralysis of the vaso-motor nerves there is dilatation of the vessels in connection with them, and so a larger amount of blood is carried to the part, and consequently a larger quantity of oxygen acts upon the tissues. But an increased supply of blood is not sufficient of itself to increase the functional activity of an organ, and therefore to produce increased heat. This has been clearly demonstrated by the well-known experiments on the submaxillary gland, and by the experiments on the muscles. The submaxillary gland is supplied by two nerves, by branches of the chorda tympani reaching it along its duct, and by branches of the cervical sympathetic reaching it along its arteries. Now the chorda contains two sets of fibres, (i) secreting fibres acting only on the epithelium cells, and (ii) vaso-motor or dilating fibres. Stimulation of this nerve brings about two events, dilatation of the blood-vessels of the gland and a flow of saliva ; that is, there is not only a dilatation of the vessels, and so increased supply of blood, but there is an absorption, an assimilation, of some of the molecular constituents of the blood by the gland cells, and if these constituents are pro-perly elaborated by the cells, the formation of saliva. But the elaboration of these constituents can be modified, for by stimu-lating the sympathetic nerve "a slight increase of flow is seen, but this soon passes off, and so much saliva as is secreted is remarkably viscid, of higher specific gravity, and richer in cor-puscles and protoplasmic lumps, and is said to be more active on starch than chorda saliva.*" This fact has a very important bear-ing upon the subject I am discussing—the fact, namely, that ac-cording to the variations in the stimulus applied to the two sets of nerves, alterations in the secretion can be produced: modifications, that is to say, in the molecular changes in the contents of the secreting cell can, in this way, be brought about.

Further, if a small quantity of atropin be injected into the veins, stimulation of the chorda produces no secretion of saliva at all, though the dilatation of the blood-vessels takes place as usual.

---

* Foster's *Physiology*, p. 261.

This proves "that atropin, while it has no effect on the vaso-
motor fibres, paralyses the secreting fibres just as it paralyses the
inhibitory fibres of the vagus.   Hence, when the chorda is
stimulated, there pass down the nerve, in addition to impulses
affecting the blood-supply, impulses affecting directly the protoplasm of the secreting cells, and calling it into action, just
as similar impulses call into action the contractility of the protoplasm of a muscular fibre.   Indeed the two things, secreting
activity and contracting activity, are quite parallel.   We know
that when a muscle contracts, its blood-vessels dilate ; and just as
by atropin the secreting action of the gland may be isolated from
the vascular dilatation, so by urari muscular contraction may
be removed, and leave dilatation of the blood-vessels as the only
effect of stimulating the muscular nerve *."

If this is the case, there must be, in order that heat may
be developed, not only an increased supply of blood to a tissue,
but also some chemical change in its molecular constituents.
Such a molecular transformation is shown by the production of
lactic acid and carbonic acid when a muscle contracts or when it
dies.   The molecules forming these substances are detached from
the muscular tissue.   Now I would suggest that the change in
the molecular constituents of the muscular tissue which leads to
the further development of heat, results from a weakening or
lessening of the power, whatever that may be, which holds the
molecules together ; that with the dilatation of the vessels in the
part, under the influence of the vaso-motor nerves, there is also
a splitting up or tumbling to pieces of the albuminoid molecules, and from both causes heat is developed.   The normal
change of the cyan-alcohols is interfered with.   Their condensation
into higher cyan-alcohols with elimination of urea†, or their
change into cyanamides, and amido-acids with the ultimate
oxidation of the latter, is modified.   The molecules $CH_2\begin{cases} OH \\ CN \end{cases}$

---

* Foster's *Physiology*, p. 261.   † See page 26.

and $C_2H_4 \begin{cases} OH \\ CN \end{cases}$* become detached more or less from each other, and by hydration (by which heat is developed) form substances— glycollic acid and lactic acid—which are readily oxidised, more readily than the amido-acid glycocine which is also formed. The oxygen, though conveyed in larger quantities than normal by the increased blood-supply to the tissue, is completely used up in oxidising into carbonic acid and water, the glycollic and lactic acids† which have been formed; the excess of lactic acid and the glycocine are unoxidised and pass into the circulation. Now, in a normal or healthy state of things, the irritating or stimulating cause acting on the vessels and nerves of the skin being removed, reaction would be set up there; the cutaneous vascular area, acted upon by the lactic acid, would dilate, and consequently the vessels of the muscular area would contract; and this latter contraction would be increased by the stimulating effect of the glycocine, or some resulting morbid product, which, being a morbid product, would stimulate some portion of the nervous system. We should naturally expect that, under the normal state of things, it would so act upon the nervous system as to check the further formation of the morbid material. This could be done by stimulating the vaso-motor nerves connected with the vessels of the muscular area, and causing their contraction. We have such an example of stimulation, though in a different direction, in the effect of carbonic acid on the respiratory centre. A closer illustration may perhaps be found in what results from the injection of bile (which contains glycocine) into the blood. This is at once followed by increased arterial tension. But the closest analogy may be found in the effects produced on the system by what is simply a form of vegetable uric acid, namely Caffeine‡. In moderate doses, this increases the heart's action both by its direct effect on the organ, and also by exciting contraction of the arteries; the blood pressure and the frequency of the pulse are both intensified.

* See page 23.   † See p. 22.
‡ Compare the formulæ for uric acid and caffeine on pages 61 and 122.

After somewhat large doses, the pulse becomes very frequent, irregular and intermittent. The morbid product then—which, as I shall endeavour to show, is uric acid—by increasing the arterial contraction, would, in a healthy state of things, cause less blood to go to the muscular area, the molecular constituents would again be held more firmly together and undergo the normal changes, the uric acid being meanwhile excreted by the kidneys, and the lactic acid by the skin. I have here drawn you a picture of an ordinary feverish cold—the primary chill, the ensuing febrile condition, the elimination of lithates with the subsidence of the attack.

The morbid product I have said, is uric acid, and results from the excessive formation of glycocine in the muscular tissue. This brings me to the questions: What relationship has glycocine to uric acid, and how are these concerned in the phenomena of rheumatism?

We know that in gout an abnormal amount of uric acid is present in the blood. The researches of Dr Garrod have shown this very clearly. But how the uric acid is formed, why in some cases it leads to uric acid calculi, why in others it passes off in the urine as gravel, in others as urate of ammonia, and in all these cases, it may be, without the joints being disturbed; why again, in other cases associated with uric acid in the blood, the joints become affected, and present the symptoms known as gouty, are points still open for discussion; and I will therefore, as I think they are closely related to, and have a distinct bearing upon, the phenomena of rheumatism, turn aside and detain you for a short time whilst I try to throw some light upon them.

Two years ago I published a paper on the formation of uric acid* in which I endeavoured to show that the formation of this substance, in the human subject, is due, in the first place, to defective assimilation or metabolism, and secondly, that it is a condensation product, much in the same way that cyanuric acid is a

* On the Formation of Uric Acid in Animals; its relation to Gout and Gravel. Cambridge, 1884.

condensation product of cyanic acid, or that cyanuric acid and biuret are condensation products of urea.

Starting with the result obtained by Strecker*, that if uric acid is heated in a sealed tube to 160°—170° C. with a cold saturated solution of hydriodic acid, it is decomposed with the formation of ammonium iodide, carbonic acid and glycocine,

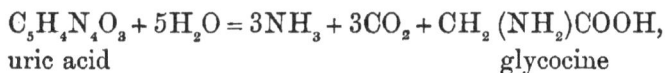

$$C_5H_4N_4O_3 + 5H_2O = 3NH_3 + 3CO_2 + CH_2(NH_2)COOH,$$
uric acid   glycocine

I endeavoured, by treating glycocine in sealed tubes and otherwise, with cyanuric acid and potassium cyanate, to reverse the process, and perform the synthesis of uric acid; but unsuccessfully. Horbaczewski, however, stated in a paper published in the *Berichte* for Nov. 1882, page 2678, that by heating glycocine and urea together, he had accomplished it. On heating glycocine with ten times its weight of urea in an open vessel to 200°—230° C. he obtained a cloudy and thick fluid. This on cooling was dissolved in weak potash solution, and supersaturated with ammonium chloride, and precipitated with an ammoniacal solution of silver and magnesia. This precipitate, containing the uric acid, was washed with water containing ammonia, and decomposed with potassium sulphide. The filtrate, freed from the precipitate, was supersaturated with hydrochloric acid, and evaporated down. On cooling, the impure uric acid separated, which was again dissolved in a weak solution of potash, and submitted twice to the same process as above. Lastly, the product was washed with alcohol and dried. It was next treated with bisulphide of carbon to dissolve the sulphur, and then treated with ether.

I tried the experiment carefully following the directions, but failed to obtain a satisfactory result. Testing the substance at various stages, I got with nitric acid and ammonia a colour somewhat approaching the murexide colour but nothing more, I failed then probably from using too high a temperature. The experiment for a long time, received no confirmation, but in May,

---

* Strecker, *Zeitschrift f. Chem.* [2] IV. 215.

1885, in the *Monatshefte für Chemie*, s. 356—362, Horbaczewski gives the following additional particulars:

" Uric acid is obtained very readily by carefully heating a mixture of glycocoll and urea in a test-tube over the flame of a small Bunsen's burner, holding the test tube in an inclined position, and moving it backwards and forwards over the flame, and taking care to keep up a continuous and free evolution of ammonia from the mass; being careful also that the temperature is not raised too high. The mass, at first clear and colourless, gradually becomes yellowish and opaque. The mass is then heated very carefully until (during the heating) it becomes quite hard, without being charred, or at least until a copious deposit is formed. The experiment is now completed. It is desirable to use only a small quantity (0·1 grm. to 0·2 grm.) of glycocoll at one time.

"In this way the whole experiment occupies only a few seconds, and the residue, after cooling, remains of a pale yellowish colour, while by using a large quantity of glycocoll, the residue appears of a dark brown colour. As regards the quantity of urea which must be used in the process, it is necessary to use a large excess. When, therefore, the glycocoll and urea are mixed together in proportion of one molecule of each, no uric acid is obtained, nor is any obtained when one molecule of glycocoll is taken with two molecules of urea. It is only when three molecules of urea are taken with one of glycocoll that uric acid is produced. It is very difficult however to isolate the uric acid in this case, and to purify it, on account of the very dark brown colour of the residue. The larger the quantity of urea taken, the more easily is the experiment made, and the residue is so much freer from colour. It is therefore advisable to mix one part of glycocoll with from 7 to 15 of urea and to heat them together in the manner above described.

"If the experiment is successful, a very small portion of the residue gives the murexide reaction; it is not so marked as with pure uric acid, but it is quite distinct.

"If the experiment has not been perfectly carried out, no murexide reaction will be obtained from the residue alone.

"The amount of uric acid obtained by this method, even when successfully performed, is rather small. In several experiments, from 1 grm. of glycocoll and the proportionate quantity of urea, only 50—150 Mlgr. of impure uric acid were obtained. The explanation of this unsatisfactory result appears to be that the reaction only takes place at a high temperature, at which also the uric acid is in some measure decomposed."

After reading this paper, I again tried the experiment, and by adopting the following method, succeeded in producing the substance.

I took thirty grains of glycocine and three hundred of urea, and put them into an ordinary test-tube (6 in. by $\frac{3}{4}$ in.). Having inserted a thermometer, to note the exact temperature, the test-tube was then placed in an oil bath, the temperature of which had been previously raised to 230° C. The bath was maintained at this temperature, and the heat so arranged that the bath could be raised in a few seconds to 240°. The mixture of glycocine and urea quickly melts; at 110° it begins to boil, and ammonia is given off; as the temperature rises it begins to froth up, and between 175° and 180° does so to a considerable extent. By raising the test-tube from the bath, from time to time, this action can be controlled, and the temperature kept at 180° for some minutes, until the ammonia comes off much less freely, and the mixture acquires a very faint yellowish tinge. The heat of the bath must then be rapidly raised to 240° and the temperature of the mixture in the test-tube quickly raised to 210°. By lifting the tube from time to time out of the bath the temperature must be kept exactly between 210° and 212° for four minutes, when the mixture will acquire the colour of brown sherry, but remains perfectly clear. The test-tube can now be removed from the bath and allowed to cool, when the mass will assume a very pale fawn colour. The mass can now be almost dissolved in three ounces of warm water, and perfect solution takes place on the addition of solution of potash. The fluid is then supersaturated with ammonium chloride, and considerable excess of ammonia added, otherwise a mixture of urate and cyanurate of potash may be precipitated. A mixture of two fluid drachms of the solution of ammonio-nitrate of silver of the British Pharmacopœia, the same quantity of the solution of ammonio-sulphate of magnesia, and of strong solution of ammonia, is now added to the other liquid. A considerable brownish precipitate of the mixed urate and cyanurate of silver is thrown down. Let this stand for twenty-four hours and allow the precipitate to subside. Decant off the supernatant liquid, and wash the precipitate with water, to which a little ammonia has been added. Take now two parts of solution of potash, and, having saturated one part with sulphuretted hydrogen, mix it with the other part. Add sufficient of this solution to decompose the argentic precipitate, heat the mixture gently, and filter from the insoluble sulphide. To the filtrate add sufficient hydrochloric acid to produce a slight but distinctly acid reaction. Boil the mixture until all the sulphuretted hydrogen is driven off, and, on cooling, uric acid mixed more or less with cyanuric acid, with very little colouring matter will be deposited. Pour off the mother liquid and evaporate it down, when a further amount of uric acid may be obtained. The presence of uric acid at this stage is shown very distinctly by the murexide test. In the earlier stages the murexide reaction is very faint indeed. In this way, though the yield is comparatively very small, I have obtained extremely well-marked crystals* presenting all the characteristics when seen under the microscope of ordinary uric acid, and responding perfectly to the murexide test†.

The following equation represents the changes

$$3CO\begin{cases}NH_2\\NH_2\end{cases} + CH_2\begin{cases}NH_2\\COOH\end{cases} = 3NH_3 + 2H_2O + C_5H_4N_4C_3$$

$$\text{urea}\qquad\qquad\text{glycocine}\qquad\qquad\qquad\qquad\text{uric acid}$$

---

* Several specimens of Uric acid, prepared by this method, were exhibited at the lecture.

† The solution of potassium sulphide must be pure. If on adding hydrochloric acid to the solution of uric acid in this, sulphurous acid is evolved, the uric acid is decomposed and not precipitated.

But it is important to know what are the intermediate stages, if we are to understand the mode of formation of this substance in the living body.

Now hydantoic acid has been formed synthetically, and here we have the first step towards the synthesis of uric acid. By heating urea and glycocine to a temperature of $120°—125°$ C. they combine and form hydantoic acid *

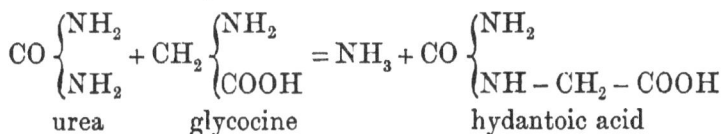

$$CO \begin{cases} NH_2 \\ NH_2 \end{cases} + CH_2 \begin{cases} NH_2 \\ COOH \end{cases} = NH_3 + CO \begin{cases} NH_2 \\ NH - CH_2 - COOH \end{cases}$$

$$\text{urea} \qquad\quad \text{glycocine} \qquad\qquad\quad \text{hydantoic acid}$$

and this dehydrated forms hydantoin; for by boiling the latter with baryta water it is converted into hydantoic acid

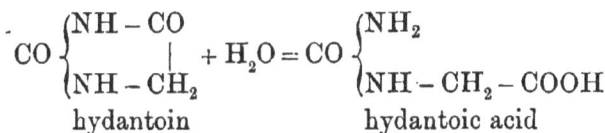

$$CO \begin{cases} NH - CO \\ \quad | \\ NH - CH_2 \end{cases} + H_2O = CO \begin{cases} NH_2 \\ NH - CH_2 - COOH \end{cases}$$

$$\text{hydantoin} \qquad\qquad\qquad\quad \text{hydantoic acid}$$

By heating urea alone to a temperature somewhat higher, viz. to $150°—160°$ C., condensation takes place, and it is converted into biuret $C_2O_2N_3H_5$.

$$2CO \begin{cases} NH_2 \\ NH_2 \end{cases} = NH_3 + \begin{matrix} CO \begin{cases} NH_2 \\ NH \end{cases} \\ CO \begin{cases} \\ NH_2 \end{cases} \end{matrix}$$

$$\text{urea} \qquad\qquad\qquad \text{biuret}$$

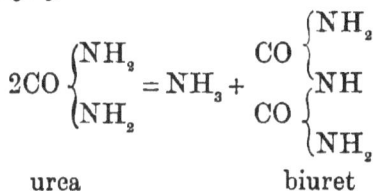

If now we combine together hydantoin and biuret we have the elements of water and ammonium urate,

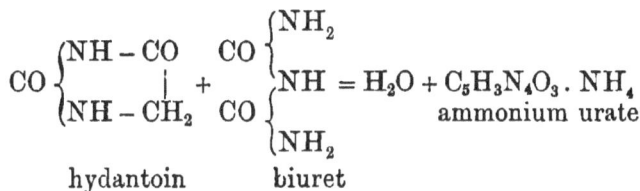

$$CO \begin{cases} NH - CO \\ \quad | \\ NH - CH_2 \end{cases} + \begin{matrix} CO \begin{cases} NH_2 \\ NH \end{cases} \\ CO \begin{cases} \\ NH_2 \end{cases} \end{matrix} = H_2O + C_5H_3N_4O_3 . NH_4$$

$$\text{hydantoin} \qquad\qquad \text{biuret} \qquad\qquad\qquad \text{ammonium urate}$$

* Heintz, *Jahresb. für Chem.* 1865, S. 360.

In performing Horbaczewski's experiment, when the temperature is raised to 120°—125° C. hydantoic acid is formed, and at 150°—160° C. the urea condenses into biuret. If then between 130° and 160° C. the hydantoic acid is dehydrated into hydantoin which melts at 207°, and at that temperature or one something above, combined with the biuret in the manner above shown, forming water and ammonium urate, we have a complete explanation of the process.

That the synthesis takes place in this way, I have confirmed in some measure by experiment. One part of hydantoin, with four parts of biuret, were treated in the same manner as in the experiment with urea and glycocine. Throughout, the same results follow as when operating upon glycocine and urea; but the yield of uric acid is smaller. On evaporating the solution after the addition of hydrochloric acid, crystals are deposited which give the murexide reaction very distinctly, but as yet I have not been able to isolate the uric acid*. The constitution of glycoluril $C_4H_6N_4O_2$ which can be obtained from uric acid and from allantoin, also favours this view. By boiling glycoluril with acids it is converted into hydantoin and urea. Consequently by combining these substances together in a suitable way we ought to perform the synthesis of glycoluril.

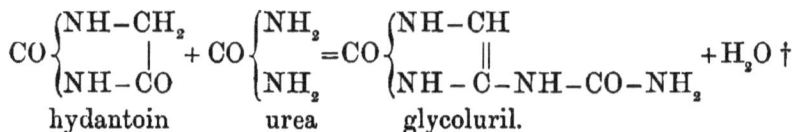

$$CO\begin{cases}NH-CH_2\\ \ \ \ \ |\\ NH-CO\end{cases} + CO\begin{cases}NH_2\\ NH_2\end{cases} = CO\begin{cases}NH-CH\\ \ \ \ \ \ ||\\ NH-C-NH-CO-NH_2\end{cases} + H_2O\ \dagger$$

hydantoin   urea   glycoluril.

If, then, hydantoin were combined with biuret instead of with urea, we should have

---

* The reason probably is that the hydantoin combines more readily with the biuret in its nascent state, that is as condensation of urea takes place, than with biuret fully formed.

† Strecker-Wislicenus, *Org. Chemistry*, London, 1881, p. 539.

$$CO \begin{cases} NH - CH_2 \\ \quad\quad | \\ NH - CO \end{cases} + \; CO \begin{cases} NH_2 \\ NH \\ NH_2 \end{cases}$$

hydantoin   biuret

$$= CO \begin{cases} NH - CH \\ \quad\quad || \\ NH - C - NH - CO - NH - CO - NH_2 \end{cases} + H_2O$$

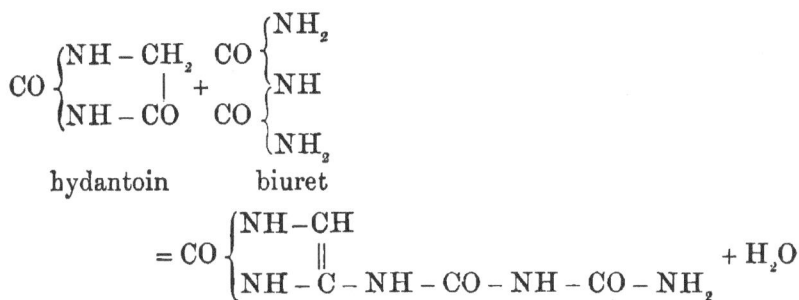

and ammonia being given off, the compound would be

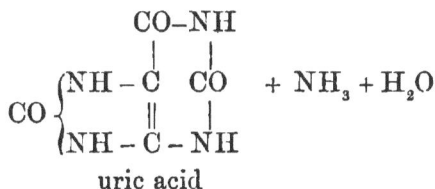

$$CO \begin{cases} \quad\quad CO-NH \\ \quad\quad | \quad\quad | \\ NH - C \quad CO \\ \quad\quad || \quad\quad | \\ NH - C - NH \end{cases} + NH_3 + H_2O$$

uric acid

which is the molecular formula for uric acid given by Medicus, and the one most generally accepted *.

My explanation of the formation of uric acid in the animal economy, based on these considerations, is as follows :

In the human subject glycocine conjugated with cholic acid is poured out as glycocholic acid, a constituent of the bile, into the intestine. After the bile has served its purpose in digestion, the glycocine as well as taurine are returned into the blood. These together with the other amido-bodies, leucin, and possibly tyrosin, the products of the digestion of albuminous food, reappear in the urine as urea ; that is, the urine does not contain them, but its urea is proportionately increased. Now these amido-bodies, glycocine, leucine, &c., are probably carried by the portal vein straight to the liver, and, from certain facts which I have stated in my first lecture, we are led to the view that among the numerous

---

* This explanation of the synthesis of uric acid indicates I think the way in which guanine, xanthine, hypoxanthine, theobromine and caffeine may each be built up from glycocine, thus showing their relationship to uric acid. The different steps in the process are indicated in the Appendix, page 120.

metabolic events which occur in the hepatic cells, the formation of urea from these bodies may be ranked as one. Suppose from some cause this metabolism of glycocine is interrupted (and I need only refer here to the interrupted metabolism of starch or glucose in diabetes as an illustration of what I mean), whilst taurine, leucine, &c., still undergo the normal changes with the production of urea, we should then have in the gland the two substances, glycocine and urea, (or the immediate antecedent of urea) the conjugation of which by the gland (just as in the case of hippuric acid being formed from the conjugation of glycocine and benzoic acid) would produce hydantoic acid,

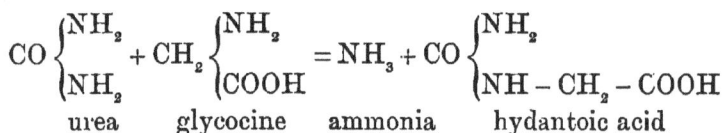

$$CO\begin{cases} NH_2 \\ NH_2 \end{cases} + CH_2\begin{cases} NH_2 \\ COOH \end{cases} = NH_3 + CO\begin{cases} NH_2 \\ NH - CH_2 - COOH \end{cases}$$

$$\qquad urea \qquad glycocine \qquad ammonia \qquad hydantoic\ acid$$

which dehydrated would be converted into hydantoin

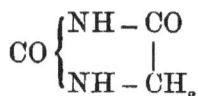

$$CO\begin{cases} NH - CO \\ \quad\quad | \\ NH - CH_2 \end{cases}$$

Hydantoin is easily soluble, and so would pass on in the circulation to be combined elsewhere with two molecules of urea or with biuret, which is also soluble, to form ammonium urate*

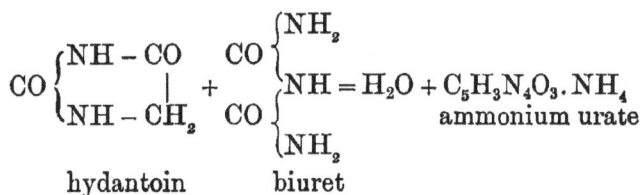

$$CO\begin{cases} NH - CO \\ \quad\quad | \\ NH - CH_2 \end{cases} + \begin{matrix} CO\begin{cases} NH_2 \\ NH \\ \end{cases} \\ CO\begin{cases} \\ NH_2 \end{cases} \end{matrix} = H_2O + C_5H_3N_4O_3.NH_4$$

$$\qquad hydantoin \qquad\qquad biuret \qquad\qquad ammonium\ urate$$

---

* In the paper published two years ago I suggested that the synthesis might take place also, in another way, by the conjugation of glycocine first with biuret, which afterwards combining with urea would form ammonium urate, but this would give a different molecular arrangement for uric acid, and the actual production of uric acid from hydantoin and biuret shows that the interpretation given in the text, is the correct one.

It may be urged as an objection, that in order to produce hydantoic acid and biuret out of the body a temperature of 160° C. is required, and therefore that the body at the normal temperature of 37°—38° C. cannot produce this result. The fact that the conjugation of benzoic acid and glycocine takes place in the body at the normal temperature, when benzoic acid is introduced into the alimentary canal, hippuric acid being formed and passing off in the urine, to effect which conjugation out of the body requires exactly the same temperature of 160° C., is a complete answer to such an argument. But I will show directly that the combination of urea, or its antecedent ammonium cyanate, with other bodies does actually take place in the animal body.

Horbaczewski, in his paper in the *Monatshefte*, also describes a method for obtaining methyl uric acid by synthesis. He says :

"It was naturally to be expected, that if in the previous experiment, instead of glycocoll a derivative of glycoll was employed, instead of uric acid a derivative of uric acid would be obtained. The experiment was performed with sarcosine and confirmed this supposition. When sarsocine is heated in small portions (0·1—0·2 grm.) with 5 to 10 parts of urea, after the method described for uric acid, and carefully warmed until the mass becomes solidified whilst hot—which can always be effected in the present case—the residue is found to contain methyl uric acid. This can be demonstrated at once in the residue by the murexide test, in fact a small atom generally gives a much more decided reaction than can be obtained from the uric acid residue. If the experiment has been successful the reaction is very striking. The methyl uric acid can be separated from the residue in exactly the same way as described for uric acid. The purification of this compound can be effected far more easily than that of uric acid, and presents no difficulties whatever."

The changes which have taken place here are, I venture to suggest, of the same character as those indicated for the formation of uric acid. The sarcosine is combined first with one molecule of urea, forming methyl hydantoic acid, which dehydrated into methyl hydantoin, and then combined with biuret, forms ammonium urate.

$$CH_2 \begin{cases} NH.CH_3 \\ COOH \end{cases} + CO \begin{cases} NH_2 \\ NH_2 \end{cases} = NH_3 + CO \begin{cases} NH_2 \\ N.CH_3 - CH_2.COOH \end{cases}$$

$$\text{sarcosine} \qquad \text{urea} \qquad\qquad \text{methyl hydantoic acid}$$

$$CO \begin{cases} NH_2 \\ N.CH_3 - CH_2 - COOH \end{cases} = CO \begin{cases} NH\text{——}CO \\ \quad\quad\;| \\ N.CH_3 - CH_2 \end{cases} + H_2O$$

$$\text{methyl hydantoic acid} \qquad\qquad \text{methyl hydantoin}$$

$$CO \begin{cases} N.CH_3 - CH_2 \quad CO \\ \qquad\qquad | \\ NH\text{——}CO \end{cases} + CO \begin{cases} NH_2 \\ NH \\ NH_2 \end{cases}$$

$$\text{methyl hydantoin} \qquad\qquad \text{biuret}$$

$$= CO \begin{cases} & CO - NH \\ & \;|\qquad\;\; | \\ N.CH_3 - C & \quad CO \\ \qquad\quad \| & \quad | \\ NH\text{——}C & - NH \end{cases} + NH_3 + H_2O$$

$$\text{methyl uric acid.}$$

Now methyl hydantoin has been actually formed after the administration of sarcosine to a living animal, and detected in the urine. Considerable discussion took place at the time this was announced, as to whether or not methyl hydantoin really appeared in the urine, and a number of experiments were made by Schultzen, &c. The importance attached to the question at the time arose from the fact, that it appeared to demonstrate the existence of carbamic acid in the animal system, and to show that this acid was the immediate precursor of urea. A good deal may be advanced in favour of this view of the formation of urea. I have however endeavoured to show that the antecedent of urea is ammonium cyanate; and as methyl hydantoin has been formed by mixing together sarcosine, potassium cyanate and ammonium sulphate, and digesting them at the temperature of the body

(40° C.), the experiment with sarcosine possesses great interest and significance, and throws very clear light upon the steps by which uric acid is formed in the animal economy. In the *Zeitschrift für physiologische Chemie* Dr J. Schiffer gives an excellent summary of the experiments and views of the other authorities, together with an account of his own observations and experiments. From this paper* I take the following extracts.

" Few investigations in the range of physiological chemistry have in recent times excited so much attention as those of Schultzen on the transformation of sarcosine in the animal body. In conjunction with Leon v. Nenki (*Zeitsch. f. Biologie*, Bd. VIII.) he had previously found on giving glycocine as food the amount of urea excreted corresponded with the amount of N given...... He repeated his experiments with methyl glycocine or sarcosine. Increased secretion of urea did not take place ; on the contrary two new bodies appeared in the urine, both having an analogous composition, the one compounded of sarcosine and carbamic acid, the other of sarcosine and sulphamic acid. The first of these bodies was identical in its constitution with methyl hydantoic acid, but was not recognized as such by Schultzen. He concluded from his investigation that the sarcosine attached to itself the carbamic acid resulting from disintegration of the albuminous bodies and which in a normal condition gives rise to the production of urea. With this hypothesis, it appeared very plausible that urea should disappear from urine containing sarcosine. He imagined therefore he had explained the mode in which urea was formed, and had thus solved one of the most important questions in physiological chemistry by the convincing proof of a carefully devised experiment."..............

"Salkowski found after administering taurine its uramido-acid tauro-carbamic acid in the urine (*Berichte*, VI. s. 744)."............

"Later E. Bauman and Hoppe-Seyler (*Berichte*, Bd. VII. s. 34) succeeded in forming methyl hydantoic acid synthetically under

* *Zeitschrift für physiol. Chemie*, Bd. v. S. 267.

such conditions as might exist in the animal body. Equivalent
amounts of sarcosine, potassium cyanate and ammonium sulphate
were digested at a temperature of 104° F., the potassium sulphate
removed by alcohol, and the baryta salt of the acid referred to
obtained."

"In a similar manner Salkowski (*Berichte*, vii. S. 116) at the
same time produced this acid, or rather, as it is easily decomposed,
its anhydride, methyl hydantoin."

"So far everything seemed to confirm Schultzen's experiments.
But when further experiments were made, essentially different
results were obtained. These experiments were undertaken by
E. Salkowski on the one hand and Bauman and v. Mering on the
other. The results obtained by all completely demonstrate the
absence of sarcosine sulphamic acid. As regards methyl hydantoic
acid Salkowski (*Berichte*, viii. S. 115) first stated that in the urine
of dogs it appeared only in small quantity after the administration
of sarcosine, whereas Bauman and v. Mering in their experiments
on the human subject showed that after administering as much
as 25 grammes of sarcosine, methyl hydantoic acid was entirely
absent from the urine, and that Schultzen in his experiments could
not have had this substance to deal with. At the same time they
discovered a probable source of error in his experiments, viz. that
in the presence of sarcosine Liebig's test for urea fails. They discovered also, as Salkowski also did, that a portion of the sarcosine
appeared unchanged in the urine. In later communications Salkowski confirmed the view of Bauman and von Mering that, after
the administration of sarcosine there is no appearance whatever of
methyl hydantoic acid in the urine."

"There appeared to be very little left then from Schultzen's
experiments. One point only still remained for investigation. As
the uramido-acids are so easily converted into their anhydrides,
and as this is specially so in the case of methyl hydantoic acid and
its conversion into methyl hydantoin, this latter substance might
possibly exist in the urine after the internal administration of

sarcosine. Salkowski has given great care and attention to this point, without however arriving at any definite proof."............

"This then was the state of the question, when Professor Bauman informed me that methyl hydantoin reduces sulphate of copper in an alkaline solution and asked me to make some fresh experiments with sarcosine founded upon this reaction."

Schiffer then describes the experiments which he performed to demonstrate the existence of methyl hydantoin in the urine, and thus sums up the result of his investigations:—

"Our knowledge therefore of the destination of sarcosine in the organism may be formulated as follows : By far the greatest part is excreted unchanged : a smaller portion, one-fifth to one-sixth, is transformed into the uramido-acid we have been discussing, or rather into its anhydride (methyl hydantoin), and a smaller portion is oxidised into methyl urea."

Sarcosine then undergoes little change in passing through the system, but a portion is converted into methyl hydantoin. If therefore when glycocine is absorbed from the alimentary canal, it does not undergo the normal change, we may reasonably suppose that when brought in contact with ammonium cyanate, it will be acted upon in the same way as sarcosine, and a portion transformed into hydantoin, and thus we have the first step towards the formation of uric acid. Circulating in the system, the hydantoin may, under certain conditions, so affect the central nervous system, and thereby the innervation of the kidneys, that when it arrives at these organs, condensation of two molecules of ammonium cyanate may take place with elimination of ammonia, forming biuret, and thus be conjugated with hydantoin ; or this condensation may take place independently of the nervous system. Condensation of the cyanates readily takes place, as is shown by the action of acetic acid on potassium cyanate, forming potassium cyanurate and potassium acetate

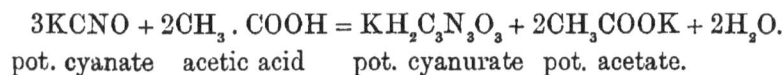

$$3KCNO + 2CH_3 . COOH = KH_2C_3N_3O_3 + 2CH_3COOK + 2H_2O.$$
pot. cyanate   acetic acid   pot. cyanurate   pot. acetate.

Such a view I think meets all the difficulties which have been
raised with regard to previous theories as to the formation of
uric acid—difficulties which have been very clearly set forth by
Dr Garrod, in his Lumleian lectures. Nor is it difficult to understand that if, as I have endeavoured to prove, the final conjugation
of two soluble bodies takes place in the kidney, forming the very
slightly soluble ammonium urate 1—2400, a portion may not be
excreted but remain in the blood, overflow as it were, and so pass
on into the circulation; a result which certainly happens when
the ureters are ligatured. The ammonium salt would then, meeting with the soda in the blood, be converted into sodium urate, the
form in which it is deposited about gouty joints.

The appearance then of uric acid in the secretion is the result
primarily of the non-transformation or metabolism of glycocine
into urea—whether that glycocine be derived from the bile poured
out into the duodenum or formed elsewhere in the body.

I now come to the question of the abnormal formation of uric
acid in the human system. Just as in diabetes the essential fault
lies in the inability of the system, either in the liver, or it may be
elsewhere, to effect the metabolism of glucose which then passes
into the circulation and is discharged by the kidneys, so, in gout or
gravel, the imperfect metabolism of glycocine is the primary and
essential defect. Unchanged it passes from the alimentary canal
or elsewhere, into the liver, there under the action of the gland it
is conjugated with urea resulting from the metabolism of the other
amido-bodies, leucine &c., and is converted into hydantoin; it then
passes on to the kidneys to be combined with other molecules of
urea or biuret, forming ammonium urate, a portion of which over-
flows into the circulation and is converted into sodium urate. It
is not difficult to understand that in persons who are addicted to
the pleasures of the table, who are fond of port and who take
little exercise, the liver should become "sluggish," that the gland
cells should, from overwork, become inactive or destroyed; or the
terminations of the nerves should, from excessive stimulus,
become somewhat paralysed, and the gland in some measure

like the submaxillary after the injection of atropine. The result would be the imperfect performance of its function, and the non-metabolism of glycocine. But it is not every toper and gourmand who developes gout, nor is every gouty man necessarily a toper or a gourmand. Further, it is certain that uric acid is also present in excess in the blood under other pathological conditions which have no connexion whatever with arthritic mischief. How are these facts to be explained ?

LECTURE III.

THE metabolic function of the liver may be interfered with in diverse ways. From the stimulating effect of food during digestion in the stomach or duodenum, two things should take place in the liver : dilatation of the vessels of the gland, and certain changes in the gland-cells—results which may be compared to those obtained from the submaxillary gland by stimulating the chorda tympani ; the metabolic events being dependent upon the absorption and transformation of certain constituents of the blood by the hepatic cells, and this action controlled by the nerve fibres in connection with these cells. But these two processes, vascular dilatation and metabolic activity, are independent of each other. The vessels may dilate; but if the terminal portions of the nerves of the secreting cells be paralysed, the blood will pass through the gland more or less unchanged ; and the same effect will be produced by paralysing the nerve either at its central origin or along its course. We know that, as regards the innervation of the tongue and the sense of taste, there are nerve filaments which respond only to particular kinds of stimuli; so it is not unreasonable to assume that distributed to the liver cells there are various nerve filaments, one set regulating the transformation of saccharine, the other the nitrogenous elements conveyed to the cells by the portal circulation; and, if either set were paralysed, we should have the metabolism of the corresponding elements interfered with. How far the secretory power of the liver cells is influenced by the vagus, is not very clear; but the vessels are under the control of the dominant vaso-motor centre which is located in the medulla oblongata, where also we have the nucleus of the vagus. If, then, there be a want of harmony between the vascular dilatation and the action of the liver cells, if blood containing the products of digestion be passing through the liver in larger quantity than the cells can act upon (whether that inaction be due to exhaustion of the cells themselves, or to exhaustion of the secretory nerves in connection with them), then there will be imperfect metabolism, and consequently the formation of uric acid in the manner which I have described. So far as the imperfect metabolism of the saccharine elements is concerned we have illustrations of what I mean in the results which take place from the "diabetic puncture", or, as shown by Dr Pavy, from section of the sympathetic filaments ascending from the superior thoracic ganglion, or from the removal of the superior cervical ganglion.

There may be also imperfect metabolism of nitrogenous material, in another way; namely, if too much be introduced into the portal vein from the alimentary canal. The portion then which is least readily acted upon (namely, the glycocine) will not be transformed, and so the formation of uric acid is promoted. We see the same thing constantly, even when the liver is in a healthy and normal condition, if it have too much work imposed upon it; that is, if more nitrogenous material be introduced into the portal vein than can be transformed in the gland. And so an occasional indulgence at the table is very generally succeeded by the appearance of urates in the urine. If the liver cells be already exhausted by long continued over-stimulation, with how much greater difficulty will the perfect metabolism of nitrogenous food be effected!

In either of these modes, then, arising from defective changes in the liver, uric acid may be formed in excess, and then eliminated, or, it may circulate as a poison in the blood. Why, in some cases, is it eliminated as urates or as uric acid in the urine, or as renal or vesical calculi, without any arthritic symptoms? Why, in others, does it develop the arthritic symptoms to which we give the name of gout?

If the kidneys are sound, as in the majority of people below
middle age, the uric acid, unless in large excess, will be excreted;
it is thrown off directly it is formed, and there is little overflow
of the product into the circulation. If the urine have a neutral
reaction, it will appear as urates of ammonia, soda, &c. If,
however, other acids, such as lactic acid, be formed at the same
time, and eliminated by the kidneys, the base will be separated,
and uric acid, as such, be excreted, and, according to the
amount, will give rise to gravel, or to renal or vesical calculi.
But gradually the kidneys may become weakened or diseased; and
then, as the blood passes through them, the acid is formed, a portion passes on, and is not eliminated. As it is an abnormal product,
we are making no violent assumption in saying that it will act as
an irritant upon some portion of the nervous system. We have
seen how distinctly the similar body, caffein, acts as a stimulant to
certain portions of the nervous system. According, therefore, to
the susceptibility or sensitiveness of different portions, will be the
outward manifestations of this irritation. Such a susceptibility
or sensitiveness of the nervous system is constantly observed in
different individuals. In some, the nerve of smell is much more
easily stimulated by certain odours, than in others. In some, again,
the nerves of taste recognise differences in flavour which are totally
imperceptible to others. In children too, and in persons in enfeebled health, how readily is pyrexia developed by irritation of
the nerves of the intestine or of the digestive glands from im-
proper diet.

If the nucleus of the vagus be the sensitive spot then gastric
uneasiness, asthma or cardiac irregularities might be developed.

Dr Buzzard has suggested that in locomotor ataxy "the frequency of the coincidence of gastric crises with the osseous lesions
gives reasonable ground for the hypothesis that the latter may
depend upon an invasion of a part of the medulla oblongata
closely adjacent to the roots of the vagi. It is only as a working
hypothesis that I make the suggestion. Is there something which
we may call provisionally a trophic centre for the osseous and

articulatory system in the immediate neighbourhood of the roots of the vagi?

"As I have suggested on a previous occasion, the discovery of such a centre would materially help us to explain the remarkable association of cardiac complications with the joint-affection of acute rheumatism, as well as the sweating characteristic of this disease, and the occasional hyperpyrexia which occurs in it. And it might also help to throw light upon the obscure pathology of arthritis deformans*".

In a subsequent paper he further adds; "In a communication brought before the Pathological Society of London, in February, 1880 (*Transactions of the Pathological Society*), I made the suggestion that the gastric crises depend upon irritation of the nuclei of the vagus by sclerosis. At that time, I had no anatomical evidence to offer in support of the hypothesis, which was based upon the paroxysmal character of the attacks, so completely in accord with that characterising the attacks of lightning pains. It appeared evident to me that if sclerosis, which, when it attacked nerves of common sensation, produced pain, came to invade the nucleus of the vagus, it might be expected to give rise to symptoms like those of the gastric crises. During the meeting of the International Medical Congress in London, Professor Pierret, of Lyons, has shewn me sections of the medulla oblongata which he has lately made from a case of tabes with gastric crises. I was greatly interested in his demonstration that the fasciculus gracilis, in immediate relation with the nuclei of the vagus, exhibited distinct sclerosis†."

Professor Roy and Dr Graham Brown have recently stated ‡ that, "In the curarized dog stimulation of one uncut vagus with an induced current, causes contraction of the bronchi of both

* "On the affection of Bones and Joints in Locomotor Ataxy, and its Association with Gastric Crises," by T. Buzzard, M.D., *British Medical Journal*, 1881, Vol. I, p. 333.

† *Transactions of the International Medical Congress*, 1881, Vol. II, p. 27.

‡ *Proceedings of the Physiological Society*, *May* 10, 1885.

lungs, the contraction being usually powerful, but to this there are exceptions. Section of one vagus usually causes a marked expansion of the bronchi of the corresponding lung, which may be preceded by a slight temporary contraction, apparently due to the stimulus of the section. Stimulation of the peripheral end of one cut vagus always causes a much more powerful contraction of the bronchi of both lungs, than when the uncut nerve is stimulated with the same strength of induced current." They have also stated that these and other experiments enable them to understand how asthma may be produced.

If portions of the spinal cord in connection with the nerves distributed to the joints be more than usually sensitive, this sensitiveness or susceptibility being either inherited or acquired, then stimulation or irritation of these portions would lead to changes in the respective joints similar to those which result from nerve wounds, or from injuries to the spine. At first, the uric acid might produce little effect; but, with repeated or constant stimulation, these portions of the spinal cord would become more susceptible to the action of the morbid products and so nutritive changes or inflammation would be developed in the joints connected therewith. Suppose for a moment that the portion of the nervous system most susceptible to the action of the uric acid is the vaso-motor centre, the dominating portion of which, as we know, is located in the medulla oblongata, what effects should we expect to follow? The most powerful vaso-motor nerves are those which act upon the blood-vessels of peripheral parts, for example, the toes, fingers, and ears, while those that act upon central parts seem to be less active*; consequently, we should expect the toes and fingers to be first affected ; and just as by stimulating the sympathetic nerve of the sublingual gland we have contraction of the vessels and altered secretion from the gland, so in the toes and fingers we should have contraction of the vessels, and the metabolism of the parts more or less modified. In addition to contrac-

* Landois and Stirling, *Physiology*, 1885, p. 893.

tion of the vessels caused by impulses affecting the blood-supply, there are impulses affecting directly the activity of the protoplasm. The constructive metabolism about the joints is stimulated, whilst the blood supply and destructive metabolism are lessened, and so there is increased development or growth about the part. Further, the blood which goes to the part, contains sodium urate; consequently, this is deposited along with the other substances, in the metabolism or inflammatory changes which take place in the tissues around the affected part.

But, whilst the vaso-motor centres may be directly stimulated by the irritant, they may be also stimulated by irritation of the efferent or sensory nerves. Stimulation of the centripetal end of a divided efferent nerve causes increased blood-pressure. Have we not here, then, an explanation of the phenomena of a gouty paroxysm? The uric acid stimulating both the sensory nerve, and, through it or independently, the more active portion of the vaso-motor centre, causes pain in the joint and contraction of the vessels; the pain increases with the continued stimulation, but, after a time the nerve is exhausted or paralysed (as we know to be the case from continued stimulation of other nerves, as, for instance, in tetanizing a muscle). The vessels now dilate, destructive metabolism is stimulated; there is relief from pain, and with the relief, another indication of paralysis of the sympathetic shows itself, namely, more or less perspiration.

A similar explanation may, I think, be given of the mode in which arthritic symptoms result sometimes from nerve wounds. The wound may be such, that the sensory or efferent fibres in the nerve are injured, whilst the vaso-motor and other fibres are intact. From the irritation set up, either by the cicatrix or otherwise, the sensory fibres are stimulated, and, through it, the vaso-motor centres in the manner I have just described as occurring in gout; but as uric acid is not the irritant, and is not circulating in excess in the blood, there is no deposit of this substance in the metabolic changes taking place in the joint.

And further, if irritative, inflammatory, or degenerative changes

take place in the spinal cord anywhere along the course of the
sensory nerves, by which the sensory nerves are stimulated, this
would lead to stimulation of the corresponding portion of the
vaso-motor centres, and so to the development of those symptoms
which we know under the name of arthritis deformans, or rheu-
matoid arthritis. In this way, we can easily understand how
dysmenorrhœa may be the starting point of this joint-affection, a
clinical fact which has been recorded by Dr Ord.

Among the peculiar local affections in connection with gout,
we have, as pointed out by Dr Graves, congestion of the lobes of
the ear, a singular affection of the teeth, which consists in an
insuperable desire to grind them ; the occurrence of *tic douloureux*
of several branches of the fifth pair ; daily paroxysms of in-
tense heat of the nose, which continues for three or four hours,
the part becoming first of a bright and then of a purplish red
colour. All these symptoms may, I think, be ascribed to irritation
in that portion of the nervous system where the roots of the fifth
nerve and the vaso-motor centre are in close proximity, namely,
in the medulla oblongata. To this spot I shall have to refer
again.

But, independently of any functional or organic change in the
liver or in its nervous connections, there is a third way in which uric
acid may be developed. And that is from the excessive formation
of glycocine in the muscular tissue,—in the manner I have already
referred to as the result of what is commonly called a feverish
cold. This glycocine, passing in the blood to some glandular
organ—the liver or spleen it may be—is there conjugated into
hydantoin, and afterwards further transformed into uric acid.
When the nervous system and the kidneys are sound, this product is eliminated by the kidneys, and the attack subsides.

But suppose the individual who is exposed to damp or cold
has been previously reduced in strength, is tired out or exhausted,
that is to say, his vaso-motor system, or some portion of it, is
in a weak state, the weakness being either developed from bodily
or mental exhaustion, or it may be, inherited. Then following

the chill, there would be the same effects produced as I have
described when referring to the effect of cold applied to the skin;
but, the vaso-motor centre being enfeebled, the nerves regulating
the vessels of the muscular area would be more completely
paralysed by the external cold; and when reaction, induced in
some degree by the lactic acid, set in on the surface of the skin
there would be less power in the muscular nerves to recover
from that paralysed condition. More than this—so far as these
vaso-motor elements are concerned—the morbid products are
still circulating in the system; and, the vaso-motor centre not
having recovered itself, the continued stimulation by glycocine or
uric acid on that exhausted centre, no longer excites it, but gives
rise to further exhaustion, and consequent further dilatation of
the vessels in the vascular area connected with it, and more complete falling asunder of the molecular elements of the tissue.

Let me illustrate what I mean by reference to the sciatic
nerve.

"Division of the sciatic nerve of a mammal causes dilatation
of the small arteries of the foot and leg. Where the condition of
the circulation can be readily examined, as, for instance, in the
hairless balls of the toes, especially when these are not pigmented,
the vessels are seen to be dilated and injected, and a thermometer
placed between the toes shows a rise of temperature amounting, it
may be, to several degrees *". "But the dilatation so caused, after
a few days, disappears; the foot on the side on which the nerve
was divided, becomes not only as cool and pale, but frequently
cooler and paler than the foot on the sound side. If the peripheral portion of the divided nerve be stimulated with an interrupted current immediately, or very shortly, after division, the
dilatation due to the division gives place to constriction; the
sciatic nerve acts then quite like the cervical sympathetic, except
perhaps that this artificial constriction cannot be maintained for
so long a time, and is very apt to be followed by increased dilata-

---

* Foster's *Physiology*, 4th edition, p. 199.

tion. If, however, the stimulation be deferred for some days, until the dilatation has given place to a returning constriction, the effect is not constriction, but dilatation; the nerve then acts, so far as its vaso-motor fibres are concerned, like a muscular nerve, and not like the cervical sympathetic*". So also with regard to the mylohyoid muscle. Section of the nerve produces dilatation, but the dilatation is transient. The vessels speedily return to their former calibre; and then it is found that stimulation, of whatever strength, of the peripheral portion of the divided nerve, brings about not constriction, but dilatation.

If then the vaso-motor fibres of the muscular nerves have been weakened, the result would be that the normal stimulating effect of uric acid on the vaso-motor elements connected with motor nerves generally would be transformed into a depressing or paralysing effect, the vessels of the muscular area would be still further dilated, the molecular constituents of the tissue would be loosened, hydration and oxidation would go on causing fresh development of heat, there would be the continuous formation of glycocine and lactic acid, the glycocine giving rise to uric acid, which, accumulating in the system, produces its deleterious effect in modifying the function of the nervous system; the lactic acid dilating the smaller arteries† and stimulating the sweat centres, passing off in part by the skin. With increased stimulation of the paralysed centre of the muscular nerves, complete dilatation of the vessels in the muscular area will take place ; the molecular constituents of the tissue now fall completely asunder, the $CH_2 \begin{cases} OH \\ CN \end{cases}$ and $C_2H_4 \begin{cases} OH \\ CN \end{cases}$ being entirely hydrated into glycollic acid and lactic acid, with the rapid production of heat, and the heat is further developed by the ready oxidation of these products. In this way, to some extent, the so-called hyperpyrexia of rheumatic fever is produced.

---

* Foster's *Physiology*, 4th edition, p. 209.
† Gaskell: *Journal of Physiology*, Vol; III, p. 18.

But, whilst uric acid affects in this way the vaso-motor fibres of muscular nerves, it also affects the nutrition of the joints. I have endeavoured to explain how it does so in gout, producing first of all contraction of the arteries and constructive metabolism, by the stimulating effect of the poison on the sensitive nerve-centres, followed by dilatation of the arteries and destructive metabolism when the centres are exhausted. In rheumatism, however, the dominant centre is exhausted, to begin with, and so we have destructive metabolism and vascular dilatation, preceded by very slight, if any, constructive metabolism and contraction of the vessels. In gout, the uric acid is the result of modified innervation of the liver, or exhaustion of the hepatic cells, and so there are non-transformation of the glycocine, and the consequent formation of uric acid. In rheumatism, the glycocine results from changes in the vascular area, and in the metabolism of the muscles; and along with its formation, there is also the formation of lactic acid, by hydration, both from the cyan-alcohol $CH_2 . CH_2 \begin{cases} OH \\ CN \end{cases}$ and the cyan-alcohol $CH_3 . CH \begin{cases} OH \\ CN \end{cases}$. The nutrition of the joint is modified, as above described, by the uric acid, but the nutrition is further modified by the presence of the lactic acid, in the blood, producing dilatation of the arterioles, more particularly of those in the cutaneous area.

In this manner then, I venture to suggest, are the characteristic changes about the joints in acute rheumatism developed.

Again, if the dominant vaso-motor centre be in an enfeebled or exhausted state, and irritative, inflammatory, or other changes take place in the spinal cord along the course of or adjacent to the sensory nerves by which these nerves are stimulated, we should have, quite independently of the presence of uric acid in the blood, destructive metabolism and vascular dilatation about the joints; the extreme effects of which are seen in Charcot's disease accompanying locomotor ataxy. Regarded in this light, Charcot's disease bears the same relationship to arthritis deformans that rheumatism does to gout.

In a similar way, also I would explain the joint-affection known as gonorrhœal rheumatism. Here we have prolonged irritation of the sensory nerves of the urethra, acting through an enfeebled or weakened vaso-motor centre; and, in this way, the nutrition of the joints in connection with that centre is modified.

I have referred to that portion of the sensory tract in the medulla oblongata in the neighbourhood of the nucleus of the vagus, the root of the fifth nerve, and the dominant vaso-motor centre, as the part which controls the nutrition of the joints, and which is more particularly affected by external cold. The following facts give support to this view and can be explained on the same hypothesis.

Cases occur in which the solitary local affection, associated with a feverish cold, is an outbreak of herpes on the lips, nose, or buccal mucous membrane, the so-called *Febris herpetica.* " We find a sharp attack of fever in a child or young person, without any local cause ; we expect the onset of some serious disease, but our anxiety is soon allayed ; in a few days, there unexpectedly appears on the lips, nose, or cheeks, a closely-packed group of vesicles, which at once removes our uncertainty, and shows us that we have to do, not with the early stage of one of the more serious fevers, but with a feverish cold, which will end favourably in a day or two.*" Here we have, as the result of " catching cold," disorder of nutrition, showing itself in parts supplied by branches of the fifth nerve, namely, the labial and nasal branches of the superior maxillary nerve.

Rheumatism not infrequently, after a short interval, follows tonsillitis. Attention was prominently called to this a few years ago, by Dr Kingston Fowler, and I have seen several instances where it has occurred. Now, we may assume either that the condition of the central nervous system in the individuals so affected

* Ziemssen's *Encyclop.*, vol. xvi, p. 241. London, 1877.

is such that slight injurious influences, such as the effect of a slight
chill, or exposure to cold, acting upon the central nervous system,
will produce an attack of quinsy, whereas more severe injurious
influences will cause rheumatic fever ; or we may assume that, in
quinsy, the continued irritation of the sensory nerves distributed to
the tonsils produces exhaustion of some portion of the dominant
vaso-motor centre; and as, in quinsy, uric acid is largely developed,
this acting on the enfeebled centre exhausts it still more and
developes the arthritic symptoms of rheumatism in the manner I
have described.   Either hypothesis will illustrate my argument.
The tonsillar nerves are derived from the fifth nerve—the middle
descending branch from Meckel's ganglion—and from the glosso-
pharyngeal portion of the eighth nerve.   This branch of the fifth,
to which I have referred, is derived from the larger root of the
nerve, which may be traced back to the lateral tract of the medulla
oblongata immediately behind the olivary body, and is connected
with the grey nucleus at the back part of the medulla between the
fasciculi teretes and the restiform columns.   The deep origin of
the glosso-pharyngeal may be traced through the fasciculi of the
lateral tract to a nucleus of grey matter at the lower part of the
floor of the fourth ventricle, external to the fasciculi teretes.
Both these nerves receive filaments from the superior cervical
ganglion of the sympathetic.

The occurrence of arthritis in the course of hemiplegia, a
point to which I referred in my second lecture*, would at first
sight appear antagonistic to a theory which ascribes the joint
symptoms to neurotic changes in the medulla oblongata.   But
it only requires a little reflection to show that it gives the strongest
support to the theory.   Following apoplexy there is descending
sclerosis of the lateral columns which proceeds slowly.   When
it shows itself in the medulla, irritation may be set up around
the degenerating portion; and this extending to the grey matter
and anterior cornua, may set up the arthritic symptoms which

* See page 47.

sometimes show themselves in from fifteen days to a month after the apoplectic seizure.

As regards megrim, Dr Edward Liveing* writes, "There can be no question then, I think, as to the frequent connexion of megrim, whether in its blind, sick, or simply hemicranial forms, with a gouty diathesis, and its occasional replacement by fits of regular gout." Some years ago I endeavoured to prove† that megrim was due to vaso-motor disturbance,—to uncontrolled action, or as I should now say, to stimulation of certain branches running from the superior cervical ganglion of the sympathetic, followed by exhaustion. Dr Haig in a recent number of the *Practitioner‡* asks the question : "May not excess of uric acid in the blood cause such vaso-motor irritation?" Certainly; and in persons who are subject to megrim, whose sympathetic ganglia have this sensitive constitution, the causes which will lead to the formation of uric acid in the system, will with tolerable certainty, develope an attack of this particular disorder.

Again, with regard to chorea. If the vaso-motor fibres proceeding from the upper cervical ganglion to form the carotid plexus have, through prolonged stimulation, become exhausted or weakened, there would be, as the result of that exhaustion, vascular dilatation and circumvascular change in the track of the middle cerebral artery. But there would also be nutritive changes in the nerve cells of the brain, and exhaustion of their metabolic powers. We should, in fact, have a similar condition produced to that of the sub-maxillary gland when the chorda tympani is stimulated after the injection of atropine; and *as far as the nerve cells* are concerned, the same result would be produced as ensues from embolism. Their co-ordinating action would be paralysed in both cases, and such changes would produce the inco-ordinated movements of chorea. Nor is it difficult to suppose that sudden shock would

---

* *On Megrim and Sick-headache*, 1873, pp. 404—5.

† *On Nervous or Sick-headache*, Cambridge, 1873.

‡ March, 1866, p. 181.

give rise both to vaso-motor paralysis, and to inhibition of the assimilating power of the nerve cells. Dr Dickinson has made out very clearly that it is in the tract of the middle cerebral artery, and in the posterior and lateral parts of the grey matter, and in the upper portions of the spinal cord that vascular dilatation and circumvascular change take place both in chorea and diabetes; and I venture to suggest that they are induced in the manner I have indicated.

Will the theory I have here advanced offer any explanation of the shifting character of the rheumatic affection? My application of the theory is as follows. Vaso-motor centres are distributed throughout the whole spinal axis. "They can be excited reflexly, but they are also controlled by the dominating centre in the medulla oblongata*." "Now this *general* vaso-motor centre in the medulla oblongata is really a *complex composite centre*, consisting of a *number* of closely aggregated centres, each of which presides over a particular vascular area†." Some of the nerve cells or some portions of this dominating centre may be more readily exhausted by the continued stimulation of an irritant circulating in the blood than the others; and after the development of the irritant (uric acid) in the system, it will of course produce exhaustion, first in these particular cells and in the subsidiary ganglia in connection with them, and so lead to dilatation in the vascular area directly under their control. The other cells in the dominating centre, and the subsidiary ganglia in connection with them, possessing a healthy tone, will be stimulated to action in the normal manner, and contraction will take place in the vascular areas connected with them. This action itself would have the effect of driving more blood to the paralysed vascular area, and intensifying the symptoms there. But with continued stimulation of these healthy centres, exhaustion will sooner or later be induced; further stimulation increases the exhaustion, and then dilatation of the vessels under

---

* Landois and Stirling's *Physiology*, London, 1885, p. 854.
† *Ibid.* p. 898.

their control will be induced ; with this dilatation in a new area, the amount of blood in the parts primarily affected will be lessened, and simultaneously will the symptoms which were associated with this increased blood supply be modified or removed.

If at any time a joint has suffered injury in any way, this joint will be more sensitive than natural, perhaps more easily fatigued ; and we can readily understand that this condition must be associated with some change in that part of the central nervous system which controls the nutrition and sensation in the joint; that this spot in the central nervous system will be more sensitive —more easily acted upon by irritating causes than other parts— that by the irritation, the nerve cells in the part will be more easily exhausted ; and therefore, in rheumatic individuals this particular joint will be the one first affected, and perhaps solely affected, in all attacks of rheumatism.

The following most interesting case, which recently came under my observation in Cambridge, illustrates, I think, very well the development of arthritic mischief from irritation of the central nervous system, the irritation being chiefly localised in the medulla oblongata, and affecting the vaso-motor centre.

J. W., aged 54, a labouring man, was working on a stack in the dusk of early morning on Dec. 24th, 1885, when he fell, dropping about eight feet, on to his left side. He walked home with pain and difficulty, a distance of over a mile ; he went to bed, lay there drowsy and without appetite, very helpless and only just able slowly to move his limbs. After remaining a week in this condition, he was carried into Addenbrooke's Hospital on Dec. 31st, 1885, under Mr Wherry, to whom I am indebted for an opportunity of seeing the patient, as well as for the notes of the case.

When admitted, he was drowsy and stupid, and answered questions in a dull manner when roused ; cutaneous sensation seemed everywhere perfect ; movements of the limbs were very slow, difficult, and painful; turning over in bed was a struggling, tedious process. The grip of the hands was very feeble, especially the right, no rigors, no sweats, no bladder symptoms.   Bowels confined.

His health had always been good, and he was working daily up to the date of his accident.

Jan. 1, 1886.   Some swelling of the left wrist and fingers, and both ankle-joints, with pain on movement, and tenderness; relieved by an opiate.   Heart sounds normal.   Evening temperature, 101·6° Fahr.

Jan. 2.   Marked swelling and redness of left wrist-joint and of the joints of the first and second fingers, with œdema, and angry red appearance of the back of the wrist.   Slight swelling also of the right wrist and finger-joints.   It was extremely tender, and looked as if it would suppurate.   Beads of sweat on left forehead; a few small herpetic vesicles are seen along the course of the left supra-orbital nerve; tears and secretion excessive in conjunctival sacs of left eye; pulse 80; considerable œdema of both ankles.   Patient is drowsy, and sleeps, except when roused by pain.   No general sweating; urine, acid.   Bowels open by calomel.   Evening temp. 101·4°.

Jan. 3.   Redness and swelling in left wrist continues, as well as in both ankles; drowsiness and stupor; muco-pus in conjunctival sacs of left eyelids, not in the right.   The knees usually drawn up in bed, of which movement he does not seem conscious; he can put them down by an effort.   Morning temp. 100·2°; evening, 101·8°.

By Dr Latham's suggestion he was cupped at the root of the neck behind, in two places, about two ounces of blood drawn; no relief that night.   Evening temp. 101·8°.

Jan. 4.   He seemed better and in less pain.   Morning temp. 100·2°; evening temp. 101·6°.

Jan. 5.   Less pain and tension in the joints; morning temp. 99·2°.   His manner still stupid and dull.   Two cups were applied near the lower cervical vertebræ, and two near the lower dorsal vertebræ, and about four ounces of blood drawn.   He was ordered to take at once five grains of calomel, and one grain opium, and a black draught the following morning if necessary. Evening temp. 101⁰.

Jan. 6. Better. Morning temp. 100⁰; wrists and ankles less swollen. Bowels open three times after the calomel. Evening temp. 101⁰. Ordered, tinct. opii M XX at bed-time.

Jan. 7. Wrists and ankles better. Complains of pain in right knee, with some tenderness, but there is no swelling. Morning temp. 100·4⁰, even. 100·8⁰. Ordered four blisters, $1\frac{1}{2} \times 1\frac{1}{2}$, to four spots along the spine, to be kept open with savine ointment. Opiate to be taken every night.

Jan. 9. The right foot and ankle more painful and swelled. The other joints better. Morning temp. 101·2⁰. Ordered a blister 4 × 3 inches over the lower dorsal vertebræ. Evening temp. 101·8⁰.

Jan. 10. Feels better ; has had a good night, has no longer the dull stupid manner, mind clear ; was ordered tinct. opii M X. three times a day; the blister to be kept open with savine ointment.

From this time, continuous improvement took place. The evening and morning temperatures fell daily until the 19th, when the evening temperature reached the normal point, and the blistered surfaces were allowed to heal.

There was some œdema about the affected joints after the redness and pain had passed away, but this entirely disappeared by the 23rd. There was still at this date a little tenderness and thickening about the left wrist. The unilateral sweating of the face was noticed from time to time.

Feb. 2. The patient's condition much improved. Is sitting up. Has lost all pain, except now and then in left wrist. There is some thickening and stiffness about the joints which were affected, and his grip is still weak, though very much stronger than before.

Dr Dyce Duckworth, in an extremely suggestive and able essay* contends that gout is a primary neurosis, and that the

* *Brain*, Vol. III, 1880, pp. 1—22.

portion of the nervous system specially involved is situated in some part of the medulla oblongata. I hope that what I have advanced will tend not only to confirm this view, but also—to use the words of Sydenham, which he quotes—to clear up and explain some of "the difficulties and refinements of the disease itself."

Having indicated the way in which the symptoms of rheumatism and gout may be produced, let me now appeal to clinical experience, and show how far the results obtained from the practical treatment of these disorders supports the view I have here advanced as to their pathology.

If uric acid is the poison, the treatment for both diseases resolves itself into this; to prevent the formation of this substance, and when formed, to promote its elimination from the system. The kidneys will in time effect the latter if we satisfy the first condition, and we may do this by seizing upon or eliminating the glycocine from the system; for as this is an essential constituent of uric acid, by removing it we prevent the further formation of that body, and so remove the irritating substance by which the action of the central nervous system is perverted.

In gout, this may be effected in some measure, by benzoic acid. When benzoic acid is swallowed or introduced into the alimentary canal of a mammal, whether herbivorous or omnivorous, it appears in the urine as hippuric acid $C_9H_9NO_3$. The urine of adult herbivorous mammals contains no uric acid, this substance being replaced by hippuric acid, which varies in quantity both according to the food of the animal, and according to the amount of work or exercise it has taken. Hippuric acid is also found though in much smaller quantity, under normal circumstances in human urine. Now hippuric acid may be decomposed, by boiling it with strong hydrochloric acid, into benzoic acid and glycocine[*]

---

[*] Watts, *Dict. of Chem.* Vol. III, 1865, p. 158.

$$CH_2 \begin{cases} NH.C_7H_6O \\ COOH \end{cases} + H_2O = C_6H_5.COOH + CH_2(NH_2)COOH$$

hippuric acid   benzoic acid   glycocine

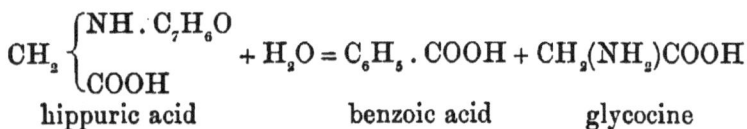

This action may be reversed; and by heating benzoic acid and glycocine in a sealed tube to 160° C. hippuric acid is formed[*]. It may also be formed by injecting benzoic acid and glycocine, or bile, into the blood of a living animal.

According to Kühne and Hallwachs[†] benzoic acid, injected alone into the jugular vein, is not converted into hippuric acid, but it is if injected into the portal vein. Meissner and Shepard[‡] however, state that after injecting benzoic acid into the jugular vein of a rabbit, at first benzoic acid alone appeared in the urine, then more and more hippuric acid, and lastly hippuric acid alone. These experimenters maintain that hippuric acid is formed in the kidneys alone, but benzoic acid appearing first alone in the urine in the above experiment would hardly bear out that view, and Salomon[‖] after introducing benzoic acid into the stomach of a nephrotomised rabbit, obtained a decided amount of hippuric acid, from the muscles, liver, and blood. Again, when benzoic acid is injected into the portal vein of some animals at least, it appears as hippuric acid in the hepatic vein[§]. Hippuric acid also appears in the urine, when benzoic acid is swallowed or introduced into the alimentary canal. Here it meets with the glycocine in the bile; and, as they pass into the liver, these are conjugated with elimination of water, forming hippuric acid,

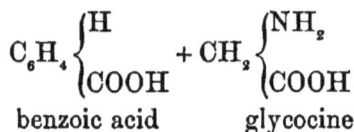

$$C_6H_4 \begin{cases} H \\ COOH \end{cases} + CH_2 \begin{cases} NH_2 \\ COOH \end{cases}$$

benzoic acid   glycocine

* Watts, *Dict. of Chem.* Vol. III, 1865, p. 156.

† *Jahresb.* 1859, S. 638.

‡ Meissner and Shepard, *Untersuchungen über das Entstehen der Hippursäure.* Hannover, 1866.

‖ W. Salomon, *Zeitschr. für phys. Chemie,* S. 365, 1879.

§ Foster's *Physiology,* 4th edition, 1883, p. 441.

$$= C_6H_4 \begin{cases} H \\ CO . NH - CH_2 - COOH \end{cases} + H_2O.$$

hippuric acid.

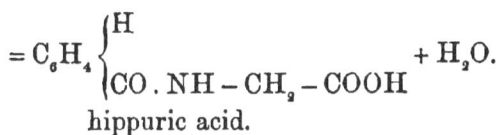

The molecule $CO . NH$, when the substance forms part of the tissue, is transformed* into $CNOH$, and then living hippuric acid would be represented by the formula

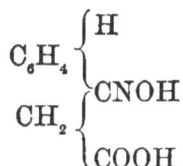

$$\begin{matrix} C_6H_4 \begin{cases} H \\ \\ \end{cases} \\ CH_2 \begin{cases} CNOH \\ \\ COOH \end{cases} \end{matrix}$$

having a constitution of the same character as that of the molecules of living albumen. But benzoic acid, after passing through the liver, may be transformed in the tissues in two ways, into hippuric acid, either by being conjugated there with glycocine, or combining with the antecedent of glycocine.

In my previous lecture, I endeavoured to show that this antecedent of glycocine is methene cyan-alcohol $CH_2 \begin{cases} OH \\ CN \end{cases}$, which, acted upon by acids, would form glycollic acid; but if first treated with ammonia, it forms a cyan-amide and then hydrated with acids or alkalis forms glycocine†.

When this substance and benzoic acid, therefore, are brought together in the tissues we should have

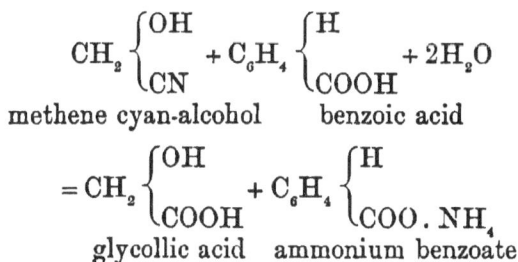

$$CH_2 \begin{cases} OH \\ CN \end{cases} + C_6H_4 \begin{cases} H \\ COOH \end{cases} + 2H_2O$$

methene cyan-alcohol   benzoic acid

$$= CH_2 \begin{cases} OH \\ COOH \end{cases} + C_6H_4 \begin{cases} H \\ COO . NH_4 \end{cases}$$

glycollic acid   ammonium benzoate

* Compare page 17.   † See page 11.

the latter of which, dehydrated, being converted into benzo-nitrile*, this becomes

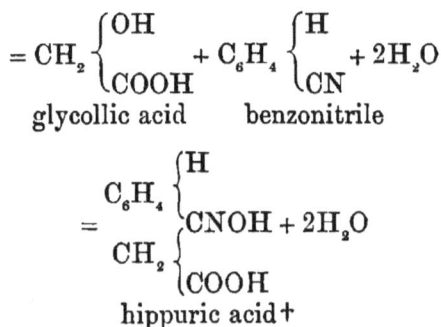

$$= CH_2 \begin{cases} OH \\ COOH \end{cases} + C_6H_4 \begin{cases} H \\ CN \end{cases} + 2H_2O$$

glycollic acid   benzonitrile

$$= \begin{matrix} C_6H_4 \\ \\ CH_2 \end{matrix} \begin{cases} H \\ CNOH + 2H_2O \\ COOH \end{cases}$$

hippuric acid †

the formula which I have just shown to be that for living hippuric acid.

It is in this way that benzoic acid acts. It seizes upon the glycocine or its antecedent, and so prevents the formation of uric acid; it passes out in the urine as hippuric acid, and gouty patients undoubtedly derive benefit from its use. Dr Golding Bird prescribed it in conjunction with phosphate and carbonate of soda, with cinnamon water as a vehicle in gout. Dr Garrod says, "I can confidently affirm that I have already obtained great advantage in the treatment of these diseases (gout and gravel) from the employment of the benzoates‡." Doubt has been expressed whether the benzoates do diminish the uric acid excretion or not. Cook‖ states that in a healthy individual the administration of benzoic acid does not stop the formation of the normal amount of uric acid, but masks its presence in the urine and interferes with the Murexide test. The later experiments of Dr Noel

* Watts, *Dict. of Chem.* Vol. i, p. 563.

† Glycollic acid and benzonitrile can, in the laboratory, be transformed into hippuric acid, by converting the benzonitrile into benzoic acid and ammonia; combining this with glycollic acid to form ammonium glycollate which, when heated, is transformed into glycocine ; and then on combining this with benzoic acid, hippuric acid is formed in the usual way.

‡ *Lancet*, April, 1883, p. 673.

‖ *Brit. Med. Journal*, July 7, 1883, p. 9.

Paton* however, clearly demonstrate "that benzoate of soda really does diminish the uric acid secretion." In an abnormal state of things, when there is an excessive amount of glycocine passing unchanged into the blood, the benzoic acid seizes upon this and converts it into hippuric acid; and so if glycocine be necessary for the formation of uric acid, the amount of the latter must be correspondingly lessened. The remedy however must be given in sufficiently large doses, doses large enough to absorb all the glycocine, and then it may not only prevent the further formation of uric acid, but, as Cook's experiments show, it may render the uric acid already existing in the blood more soluble, and therefore more readily eliminated by the kidneys.

Salicylic acid, or oxybenzoic acid $C_6H_4 \begin{cases} OH \\ COOH \end{cases}$ is another remedy which acts like benzoic acid, and this also has been found by Dr Noel Paton† to diminish the excretion of uric acid. When administered internally, it passes off by the urine as salicyluric acid—that is, it combines, in its passage through the system, either with glycocine or its antecedent, for on treating salicyluric acid with fuming hydrochloric acid, it is resolved into salicylic acid and glycocine.

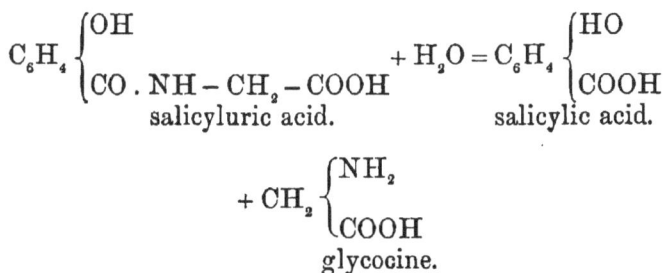

$$C_6H_4 \begin{cases} OH \\ CO.NH-CH_2-COOH \end{cases} + H_2O = C_6H_4 \begin{cases} HO \\ COOH \end{cases}$$
salicyluric acid. \hspace{2cm} salicylic acid.

$$+ CH_2 \begin{cases} NH_2 \\ COOH \end{cases}$$
glycocine.

Consequently, in the system, by seizing either upon glycocine or its antecedent, salicylic acid takes away an essential constituent of uric acid, and so prevents the formation of this body.

---

* *Journal of Anatomy and Physiology*, Jan. 1886, pp. 26—32.
† *Loc. cit.*, p. 25.

As the salicyluric acid passes into the living system, the CO . NH would be transformed into CNOH, and the above formula for salicyluric acid would become

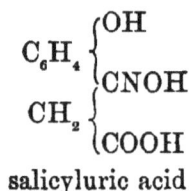

$$C_6H_4 \begin{cases} OH \\ CNOH \end{cases}$$
$$CH_2 \begin{cases} \\ COOH \end{cases}$$

salicyluric acid

or, if salicylic acid combines with methene cyan-alcohol, the antecedent of glycocine in the tissues, it would then undergo similar changes to those I have described for benzoic acid.

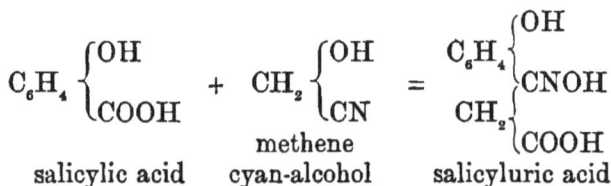

$$C_6H_4 \begin{cases} OH \\ COOH \end{cases} + CH_2 \begin{cases} OH \\ CN \end{cases} = C_6H_4 \begin{cases} OH \\ CNOH \end{cases} CH_2 \begin{cases} \\ COOH \end{cases}$$

salicylic acid   methene   salicyluric acid
    cyan-alcohol

In gout, uncomplicated with contracted kidney or albuminuria*, salicylic acid is often of service†. But it is in acute rheumatism that it shows its special power, acting truly, when properly administered, as a distinct specific. Here is a disorder, which under different treatment, may exist for weeks stationary, so to speak, in its intensity, the great heat and nervous and vascular excitement, and pain and swelling exactly of the same amount to-day as they were weeks ago; a disorder which, less than fifty years ago was said to be "often such in itself, and such in its appalling incidents, as to need, from time to time, that medicine should put forth the full compass of all its powers. Every organ or system of organs, which either directly or indirectly can receive the impression of remedies, are from time to time called to bear all that they can

---

* Paton's experiments show "that this drug has really an irritating action on the kidneys, loc. cit., p. 25.

† See Sir Dyce Duckworth's remarks on the use of salicylate of soda in gout. Year Book of Treatment for 1884, p. 81.

possibly endure; and it is often only when the powers of medicine are pressed even to the verge of destroying life that life is saved *."

And now, with or without the administration of a purgative, as the occasion requires, the patient is placed fully under the influence of salicylic acid, and in from forty to sixty hours, not unfrequently in a shorter time, the pains in the joints have subsided, the limbs can be freely moved, and the bodily temperature has reached the normal condition. But more than this—and here the remedy shows its signal power,—in no case of rheumatism that has come under my care during the last six years, either in hospital or in private practice, has there been developed, where the heart was previously sound, any cardiac complication, such as endocarditis or pericarditis. If this can be maintained and ensured, we have indeed, in our hands, a most potent remedy. Cardiac complications constitute the chief danger of acute rheumatism, and the danger, if the disease is taken in hand soon enough, may with our new remedy be averted.

But certain conditions must be observed to ensure success in the administration of the remedy. They are as follow :

First, the true salicylic acid obtained from the vegetable kingdom must alone be employed. If you have to give large doses, avoid giving the artificial product obtained from carbolic acid, however much it may have been dialysed and purified. An impure acid will very quickly produce symptoms closely resembling delirium tremens.

Secondly, give the acid without any alkali or base. A very good form is to mix 100 grains with 15 of acacia powder and a little mucilage. Allow the mass to stand and harden, and then divide into 30 pills.

Thirdly, place the patient fully under the influence of the drug—that is, let him have sufficient to produce cerebral disturbance—i. e., buzzing in the ears or headache, or slight deafness; with the development of these symptoms, the temperature and the

---

* Dr P. M. Latham's *Works*, New Sydenham Society, Vol. I. p. 112.

pain in the joints will begin to decline. To an adult, I generally administer three doses of 20 grains (six pills), at intervals of an hour, and if the head remain unaffected, a fourth dose at the end of another hour; and then repeat the 20 grains every four hours, until the physiological effect of the remedy shows itself. In the majority of cases from 80 to 100 grains are enough. In severe cases 140 to 150 may be required. Afterwards, about 80 grains a day are sufficient, and as the temperature declines, smaller quantities will develop their physiological effects, 60 or even 50 grains being then sufficient to produce cerebral disturbance. It would appear that as long as the rheumatic poison is circulating in the system, the physiological effect—that is, the effect it produces in the healthy organism—does not show itself; acting as an antidote, the greater the amount of poison, the larger must be the dose of the remedy; but as soon as the formation of the *materies morbi* is stopped, then the excess of the remedy acts as it would in the healthy organism and its peculiar physiological effects are developed. It is a very striking illustration of the difference between the therapeutic effect of a remedy and its physiological action.

Fourthly, give the patient from 40 to 80 grains daily for ten days, after all pain and pyrexia have passed away.

Fifthly, let the patient's diet consist entirely of milk and farinaceous food for at least a week after the evening temperature has been normal. On the other hand, if the patient has meat and soup, you may look forward with fair probability to a relapse.

Sixthly, take care to maintain a daily and complete action of the bowels. Calomel is the best purgative, from 2 to 5 grains at night, followed in the morning, if necessary, with a saline draught. This is the most important adjuvant to the action of salicylic acid, and I will presently explain to you why this is the case.

Seventhly, let the patient be enveloped in a light blanket, and with no more bedclothes than are sufficient to keep him from feeling cold. The object of the treatment now is to cool the patient,—not, as in former times, to sweat the poison out of him; and the cooler he is kept the sooner will the temperature be lowered. In fever increased heat increases the metabolism, just as in a cold-blooded animal*.

These are the seven rules upon which I act. I have given the true salicylic acid, where there have been both aortic and mitral mischief; and I have also given it in rheumatism complicated with pericarditis, and as yet I have seen no bad result from it. Of course, in cases of pericarditis, accompanied with delirium, the use of the remedy requires caution; you cannot tell when the system is saturated with the remedy, and you must therefore trust to smaller doses and other means for controlling the disease. Further, if pericarditis or endocarditis, pneumonia, or pleurisy, have been developed, the remedy is powerless over the mischief which is done; it will neutralise the poison producing the mischief, so as to stop its extension; but the inflammatory exudations will undergo their usual changes unabbreviated in their course. We see the same thing in tonsillitis. Given early enough, salicylic acid will stop the mischief; but if exudation of lymph have taken place, salicylic acid is powerless to cause its absorption.

There is another important point to be noticed here. In my first lecture I showed that, from condensation of two molecules of $CH_2 \begin{cases} OH \\ CN \end{cases}$, lactic acid and carbonic acid could be formed†. By seizing therefore upon this antecedent of glycocine, salicylic acid lessens the formation both of uric acid and of lactic acid.

During the administration of even large doses of salicylic acid in rheumatism, a certain amount of uric acid is still sometimes excreted by the kidneys. How is this to be explained ?

As the molecular constituents of albumen fall asunder, the molecule $CH_2 \begin{cases} CNOH \\ CNOH \end{cases}$ may become detached, and as it passes from

---

* Stirling and Landois, *Physiology*, p. 451.
† See p. 27.

the living tissue, the CNOH would be transformed into CO – NH, consequently the above molecule would be converted into hydan-

toin $CH_2 \begin{cases} CO - NH \\ | \\ NH - CO \end{cases}$ which, having little affinity for salicylic acid,

would pass on to the glands, and conjugated with urea or biuret pass out of the system as uric acid.

I have referred to the necessity of keeping up daily and sufficient action of the bowels. The benefits resulting generally, in rheumatism, from the so-called purgative plan of treatment, have always been recognised by the older physicians as striking and satisfactory. By the judicious use of cholagogue purgatives, we eliminate the bile from the intestines, and so remove from the system a quantity of glycocine, which if re-absorbed, would lead to the consequent formation of uric acid. Calomel is unquestionably of service here. Doubts may exist as to whether it promotes the flow of bile from the liver or not; but when the bile gets into the intestine calomel will cause its evacuation. " The conclusion seems inevitable, that mercurial purgatives given to healthy persons cause the escape of large quantities of bile from the alimentary canal*." Referring to the three modes of treatment—by venæsection, by opium, and by purgatives—which were in vogue at the time, Dr P. M. Latham says with regard to the last: " As this plan of treatment works prosperously day after day in its immediate effects, so day after day it gives an earnest of the remedial impression it is exercising upon the whole disease. It abates the fever, it softens the pulse, it reduces the swelling, and it lessens the pain. In short, it subdues the vascular system like a bleeding, and pacifies the nervous system like an opiate; and often, in the course of a week, the acute rheumatism is gone. In three days there is often a signal mitigation of all the symptoms; and in a week I have often seen patients, who have been carried helpless into the hospital, and shrieking at the least jar or touch

---

* H. C. Wood, Junr. *Treatise on Therapeutics*, 2nd ed., 1877, p. 435.

or movement of their limbs, risen from their beds, and walking about the ward quite free from pain.

" Of this plan, often so striking in its operation, and often so satisfactory in its results, I have some further remarks to make. It is called the purgative plan; yet its purpose is achieved by calomel and purgatives conjointly. The purgatives would not answer the end without the calomel; of that I am quite certain; neither would the calomel answer without the purgatives, unless it produced of itself ample evacuations from the bowels. It is probable, in short, that the remedial efficacy of the plan resides essentially in the calomel; in calomel, however, not as *mercury*, but as itself—*calomel*. If the specific effect of mercury—salivation —arise, it is not only beside our purpose, and against our wish, but it begets a serious hindrance to the use of calomel in sufficient quantity for the end in view. Thus the whole plan is frustrated. Having begun one plan of treatment, we are obliged to take up with another. Time is lost, the case is perplexed, the disease is prolonged, and the patient perhaps injured."

            *   *   *   *   *   *

"Now, if in the treatment of acute rheumatism, you were to choose one indication and abide by it, and were to trust one class of remedies and to it only, you will find more cases that admit of a readier cure by the method now described, than by either of the two former. You would find the aggregate of morbid actions and sufferings, which constitute the disease, more surely reached and counteracted, and more quickly abolished by medicines operating upon the abdominal viscera only, than by those which influence either the blood vessels only, or the nerves only. You would find in calomel and purgatives a better remedy than either venæsection or in opium *."

In the earlier attacks of gout too, I have often seen marked relief follow the administration of calomel and a saline cathartic. Where there is high arterial tension, as in the gouty paroxysm, this

* Dr P. M. Latham's *Works*. New Syden. Soc., Vol. i, p. 123. 124.

may be distinctly lowered by these remedies. They check the
further formation of uric acid, which is stimulating the vaso-motor
centre, and causing the increased arterial tension, by eliminating
the bile from the intestines. In some persons calomel has a depressing effect, and when the kidneys are unsound is injurious in
its action. Where calomel is inadmissible a gentle laxative such
as rhubarb is often of service. When the object is simply to
unload the bowels in a debilitated subject, it is the best purgative.
It is said to act chiefly by increasing the peristaltic action of the
bowels throughout their entire extent, but especially that of the
duodenum. According to Rutherford, it is a cholagogue. Sir
Henry Halford recommended it as a prophylactic remedy against
gout, giving a few grains of rhubarb with double the quantity of
magnesia every day : or some light bitter infusion with tincture of
rhubarb, and about fifteen grains of bicarbonate of potash.

The saline cathartics probably act only by causing serous evacuations, and in that way carry off from the blood some of the
poison contained in it. They may also act beneficially, perhaps, by
relieving a congested liver.

The diet is another important point to be attended to in the
treatment both of gout and of rheumatism. It should be simple and
nutritious, jellies and food containing gelatine should be avoided,
as this substance furnishes glycocine. Animal food will not,
itself, produce uric acid in a healthy system, as is shown in its
absence in the urine of the carnivora, but from all kinds of meat
a certain amount of glycocine will be produced, and even if all the
rest of the nitrogenous portion, after being absorbed into the
system, were converted into urea, this would necessitate an increased elimination of urea, and consequently a greater tax on the
powers of the kidneys. If these powers be weakened there will
be, with an increased call upon the organs, less power to act; and
not only would the urea but still more the uric acid, accumulate in
the blood. The striking benefit and increased urinary secretion,
which result in some forms of albuminuria from a skim-milk
diet, that is, the simplest of all diets, very well illustrate what I

mean here. The presence of urea in the blood may, by its action on the nerve-centres, determine an increased blood supply to the kidneys, and so, in a healthy state of things, an increased secretion; but if the secretory portion be damaged, or the nerve force controlling it defective, an increased flow of blood to the part, producing a congested condition of the organ, would not expedite, it would rather hinder, the work of the secretory portion. The simpler the diet then, the less tax there will be upon the kidneys, and the better they will do their work. Let the diet then be chiefly farinaceous, with just sufficient nitrogenous food to satisfy the wants of the system; and in the acute attacks let that be in the form of milk, diluted even, if necessary.

The explanation which I have here offered of the symptoms of a gouty paroxysm, helps us to understand the beneficial action of colchicum in this disorder. According to Dr Lauder Brunton * this drug paralyses the sensory nerves, the motor nerves and muscles being unaffected. It seems to act best when the bowels are previously acted upon. If in gout, then, uric acid during the paroxysm be stimulating the sensory nerves, and, through them, the more active portion of the vaso-motor centre, and we paralyse the sensory nerves with colchicum, the uric acid no longer produces its effect, and the paroxysm ceases; but colchicum has no effect in preventing the formation of uric acid, and after the paroxysm, we must endeavour to prevent its recurrence, by putting a stop to the formation of the poison, which may be done by eliminating the bile from the intestines by mercurial or other purgatives, by giving benzoic or salicylic acid and by suitable diet. In large doses colchicum will cause purging, but marked symptoms of collapse supervene, so that it is not safe to administer the remedy in this way†.

* *Pharmacology*, 1885, pp. 968, 9.

† From what I have suggested in these lectures, it is easy to understand why, in acute rheumatism, a number of blisters, applied near the affected joints, may often cure the disease; namely, by reflex action on the vessels of the muscular area, causing them to contract, and so diminishing the metabolism and lessening the formation of glycocine.

It remains for me to say a very few words with regard to the pathology of diabetes, and to explain why I have classed it together with gout and rheumatism. If the function of the liver be interfered with, so that there is imperfect metabolism of glucose as it passes through the organ, this would be a satisfactory explanation of the origin of the disease, and we should expect in such cases that the urgency of some of the symptoms would be lessened by careful diet, abstention from saccharine and starchy food. But there are other cases in which the diet seems to have much less effect in controlling the symptoms ; it is this form that I wish briefly to discuss.

I have endeavoured to show that in acute rheumatism by the separation of the cyan-alcohols $CH_2 \begin{cases} OH \\ CN \end{cases}$ and $C_2H_4 \begin{cases} OH \\ CN \end{cases}$ from the rest of the albuminous chain, we have glycocine, and glycollic and lactic acids formed; the glycollic acid being oxidised into CO_2 and water, the lactic acid in some measure being oxidised into these products, and in some measure passing off by the skin. But suppose that, whilst the vaso-motor fibres of the muscular nerve are paralysed and the vessels dilated, the molecules $CH_2 \begin{cases} OH \\ CN \end{cases}$ are detached and hydrated into glycollic acid, but only partially oxidised, the result would be that the glycollic acid would be transformed into carbonic acid, methyl aldehyde and water,

$$CH_2 \begin{cases} OH \\ \\ COOH \end{cases} + O = CO_2 + H \cdot CHO + H_2O$$
$$\text{glycollic acid} \qquad\qquad\qquad \text{methyl aldehyde}$$

Condensation of six molecules of the aldehyde may then take place, as in plants, forming glucose

$$6H \cdot CHO = C_6H_{12}O_6$$
$$\text{methyl aldehyde} \quad \text{glucose.}$$

By urari, as I have already stated, we can put a stop to muscular contraction, that is, to the oxidation of the muscular elements, and to the formation of CO_2 when the muscular nerve is stimulated. If then the tissue be, so to speak, partially urarised, the aldehyde is not oxidised, but condenses into glucose. In urari poisoning, sugar appears in the urine, though "the exact way in which this form of diabetes is brought about has not yet been clearly made out [*]."

Let me carry you one step further in the comparison between rheumatic fever and diabetes. If in rheumatic fever the central part of the nervous system connected with the muscular nerves be so enfeebled that there is complete dilatation of the vessels in the muscular area, and a falling asunder not only of the molecules $CH_2 \begin{cases} OH \\ CN \end{cases}$, but also of the molecules $C_2H_4 \begin{cases} OH \\ CN \end{cases}$, and $C_3H_6 \begin{cases} OH \\ CN \end{cases}$, then, by the hydration and oxidisation of these hyperpyrexia would, in some measure as I have suggested, be developed. But if the complete oxidation of these molecules were interfered with, the $CH_2 \begin{cases} OH \\ CN \end{cases}$ would give rise in the manner above indicated to glucose. By hydration, $C_2H_4 \begin{cases} OH \\ CN \end{cases}$ would be converted into lactic acid, which if not completely oxidised into carbonic acid and water, would be first oxidised into carbonic acid and aldehyde

$$\underset{\text{lactic acid}}{C_2H_4 \begin{cases} OH \\ COOH \end{cases}} + O = \underset{\text{aldehyde}}{CH_3 . CHO} + CO_2 + H_2O$$

the aldehyde by condensation forming para-aldehyde, a remedy recently introduced as a hypnotic. Hence the commencing drowsiness in some stages of diabetes.

If the molecules fall still more completely asunder, and the molecule $C_3H_6 \begin{cases} OH \\ CN \end{cases}$ become detached and hydrated into oxybutyric

* Foster's *Physiology*, 4th edition, p. 425.

acid, but only partially oxidised, then from one form of this acid, acetone would be formed

$$C_3H_6 \begin{cases} OH \\ \\ COOH \end{cases} + O = CO_2 + H_2O + CO\,(CH_3)_2$$

oxyisobutyric acid   acetone

which appears in the urine towards the termination of the disease. So, then, in hyperpyrexia, there is detachment of the molecules from the benzene nucleus, with their hydration and rapid oxidation; whereas, in acetonæmia, there are detachment and hydration, but imperfect oxidation of the molecules. That is to say (if this theory is correct, and such experiments were possible), by urarising an individual suffering from rheumatism with pyrexia, the urine would become saccharine; if hyperpyrexia were present, by urari the disorder would be transformed into acetonæmia.

Now, you will find that, in some forms of diabetes, salicylic acid is of the greatest service; whereas, in others, no good results from its use.

Given in doses of from ten to twenty grains three times a day, I have frequently seen it produce marked improvement; and Dr Holden, of Sudbury, has shown me notes of cases, about to be published*, where rapid amelioration of the symptoms has resulted from its use.

The urine in these cases contains often, in addition to glucose, an excess of uric acid, and the patients suffer from neuralgic pains in the joints and limbs. It also, not infrequently, contains some substance which dissolves cuprous oxide, and so more or less interferes with the application of Fehling's test. What this is, has not yet been determined. My friend, Mr Pattison Muir, kindly examined some specimens, and made out that it is some substance which readily dissolves calcic phosphate. Possibly it may be glycollic or lactic acid. If further examination should

* See *British Med. Journal* May 1st, 1886, p. 816.

prove this to be the case, it would go a long way to support my view of the origin of this form of diabetes.

I have thus endeavoured to indicate some of the changes in the nervous system, the blood, and the tissues, which may take place in diabetes, rheumatism, and gout, and to enlarge upon the text furnished by a far seeing pathologist*, when he wrote, "Disturbance in the nervous system, in some part and form may be regarded as a factor in every case of gout. There are reasons enough for thinking that changes in the nervous centre determine the locality of each gouty process, while changes in the blood and tissues determine its method and effects; and that thus we may explain the symmetries of disease in gout—sometimes bilateral, sometimes antero-posterior—and thus its metastases."

My task is done. It only remains for me to thank you, Mr President, and the Censors, for giving me an opportunity of placing my views before the College; and to thank the distinguished Fellows and Members who have listened to me. I have brought together a number of facts, and endeavoured to draw certain inferences from them. The inferences may be wrong, but the facts remain; and I trust that in this way, at least, I may have helped to a better understanding of these disorders.

* Sir James Paget. *Clinical Lectures and Essays*, 1879, p. 382.

APPENDIX.

APPENDIX.

NOTE TO PAGE 39.

"On a New Practice in Acute and Chronic Rheumatism." By
J. K. MITCHELL, M.D., *American Journal of the Medical Sciences*,
1831. Vol. VIII. p. 55.

"IN the autumn of 1827 a patient affected with caries of the spine
was suddenly attacked with all the usual symptoms of acute rheumatism
of the lower extremities. One ankle, and the knee of the opposite leg
tumefied, red, hot and painful, afforded as fair a specimen of that disease
in its acute state as is usually met with. The usual treatment by
leeches, purgatives and cooling diaphoretics, with evaporating lotions,
had the effect of transferring the symptoms to the other ankle and
knee, and finally to the hip. Disappointed in the treatment, I began
to suspect that the cause of the irritation might lie in the affected
spine. The difficulty of cure, the transfer of irritation from one part
of the lower extremities to another, without any sensible diminution of
disease, and the fact of the existence of caries in the lumbar vertebræ,
which lie near the origin of the nerves of the lower extremities, ren-
dered probable the opinion that in the spinal marrow lay the cause of
this apparently indomitable and migratory inflammation. Under this
impression, I caused leeches to be applied to the lumbar curve, and
followed them by a blister, placed on the same spot. Relief promptly
followed these remedies, and the pain ceasing to be felt in the limbs,
was perceived only in the immediate vicinity of the spinal curve. After
the blistered surface recovered its cuticle, a few leeches placed over the

diseased spine removed the pain, and left the patient in the usual state
of indifferent health attendant on such forms of spinal disease.

Striking as were the benefits of the applications made to the spine
in a case of apparent inflammatory rheumatism, they did not lead my
mind at the time to the general conclusions which, viewing the case as
I do now, they ought to have suggested.

In the beginning of the ensuing winter another case of a similar
kind presented itself. A little female patient, having curvature of the
cervical vertebræ, was attacked in the night with severe pain in the
wrist, attended with redness, tumefaction and heat. As on the appearance of these symptoms the pain in the neck, to which she was accus-
tomed, subsided, I easily persuaded myself of the spinal origin of this
inflammation, and accordingly applied leeches to the cervical spine,
with the effect of procuring a prompt solution of the disease of the
wrist.

This case led me very naturally to the reflection that perhaps other
cases of rheumatism might have an origination in the medulla spinalis,
and depend on an irritation of that important organ.

In the following spring an opportunity of testing by practice the
truth of this opinion presented itself. William Curran, a respectable
livery stable keeper in Marshall's Court, had been for upwards of two
years afflicted with a rheumatism of the lower extremities, which
gradually deprived him of the use of his limbs, and finally confined him
to his chamber. Regular medical aid and many empirical remedies
had been procured, without an abatement of the pain, which became at
length almost intolerable. On my first visit I found him in his room,
in a paroxysm of pain. His legs were swollen from knee to ankle, and
the enlargement of the periosteum and integuments gave to the anterior
face of the tibia an unnatural prominence. In that place the pain and
tenderness on pressure were particularly developed. He was also suffering severely from pain in the scalp, which had existed for a short
time previously, and was at length almost intolerable. Along with
these symptoms appeared the usual febrile action with its concomitants.
Notwithstanding the significant hints given by the spine-cases referred
to, I treated this case for a time in the usual manner—depleted freely,
purged actively, blistered the head, and, having caused an abatement of
fever, administered corrosive sublimate and decoction of sarsaparilla.

Defeated in all my efforts, I at length suggested to my patient the
possibility that his disease was so unmanageable because we had not

applied our remedies to the true seat of disease, and that by addressing
our measures to the spine success might yet be found. Accordingly,
on the 16th of February, 1828, nine days after my first visit, I had him
cupped at the back of the neck, and as he could not bear any more
direct depletion, inserted a large seton over the lumbar spinal region.
The cupping, followed by blisters to the back of the neck, relieved his
head, and as soon as the seton began to suppurate freely, his legs
became more comfortable. From the 25th, nine days after the insertion
of the seton, I visited him but seldom, although I had seen him once or
twice a day until that period. Indeed, I paid him but seven visits after
the 25th. The last was on the 30th of March. Soon afterwards he re-
sumed his usual pursuits, and about the beginning of June the seton
was removed. Since that time he has not had a return of his com-
plaint, and is, at the date of this paper, in the full and vigorous exercise
of all his physical faculties.

I could scarcely doubt as to the cause of cure in this case, because
the treatment applied to the spine was that alone which had not been
already fully and fairly tried, either by me or those who had preceded
me. Indeed, the last applications were made with some hope of success,
and the grounds of that hope were expressed to the patient, who was
fully persuaded that the spinal treatment was the chief, if not the sole,
agent of restoration.

No other well-marked case of rheumatism presented itself in my
private circle of practice, until in the winter of 1830 Mr Teale's work
on neuralgic diseases reached this country, and began to attract towards
the spinal marrow a greater share of medical attention. Although in
his essay I found nothing directly calculated to sustain me in the opinion
I felt disposed to adopt concerning the spinal origin of rheumatism, I
rose from its perusal with increased confidence in that opinion, and
resolved to experimentally examine its truth. The first well-marked
case of simple inflammatory rheumatism which subsequently presented
itself was the following: Robert Gordon, well known as the carrier of
Poulson's *Daily Advertiser*, fifty-six years of age, of vigorous constitution
and active habits, was the subject of the attack. Observing a severe
pain in his right heel and ankle, immediately followed by redness, heat
and tumefaction, he caused himself to be largely bled, and took some
salts and magnesia. On the following day the pain and swelling increased, and the ankle and knee of the opposite limb becoming similarly
affected, he was confined to bed.

On the third day my first visit was made. The patient had then a full, strong, frequent pulse, flushed face, dry skin, whitened tongue, and complained much of the severity of the pain in his legs, and of his incapacity of enduring the slightest pressure or motion. As he had already been purged, and had used a lotion, I directed the application of seventeen cups to the lumbar region, so as to abstract twelve or sixteen ounces of blood.

Next morning found the pain almost entirely gone, does not complain of moderate pressure, and is able to move his legs without inconvenience. Ordered a draught of salts and magnesia, with an evaporating lotion of camphor in alcohol.

3rd day. Pain in legs scarcely perceptible, but the shoulders, elbows and wrists are beginning to exhibit marks of severe inflammation, expressed by pain, tumefaction, heat and redness. Ordered twelve cups to the cervical spine.

4th day. The patient sits up, complains of stiffness, but no pain except in one wrist, and that very slight. Directed Epsom salt and magnesia.

5th day. Finding nothing for which to prescribe, arranged the patient's diet, recommended the occasional use of aperients, and took leave of the case.

Called on the 10th to inquire into results, and found that there had been no return of disease.

Since that time a very severe winter has passed, during which the subject of this report has continued in his customary health, and in the pursuit of his usual employments.

The reader will, in the above case, perceive that the general bleeding, though very copious, proved of no service, and that the large local depletion of the lumbar region, benefited solely that part of the disease which lay at the peripheral extremities of the nerves, supplied by the lower end of the spinal marrow. The inflammation in the upper extremities continued afterwards in progress, and was arrested only when cups were placed over the cervical end of the spinal column. The whole case exhibits a fine exemplification of the difference in the character and extent of the influence of general and topical depletion, and proves that local blood-letting is most potent when applied to that part of the spine, which supplies with nerves the parts in a state of active inflammation.

The following case, reported by Dr Thomas Stewardson, is peculiarly interesting, because of its evident dependence on irritation of nervous masses, and the immediate and perfect remedial action of the local applications.

CASE I.—William Anderson, coloured man, a seaman, aged fifty, was admitted into the Pennsylvania Hospital on the 31st of December, for a chronic rheumatism of upwards of five years duration.

Occasionally the disease intermitted, but generally continued to affect him during the cold season. The pain affected, at one or at various times, almost every part of his *right* side from head to heel, but had in no case at any period, crossed to the opposite side. Like other cases of chronic rheumatism, it was most severe in cold weather, and when warm in bed. According to his statement he seldom suffered from a winter attack for a less period than three or four months, and the existing exacerbation had lasted only a few weeks.

On the 2nd of January, two days after his admission, eight cups were applied to the *back of the neck* and *left side of the head,* and a powder was taken, consisting of guaiacum and nitrate of potassa, of each ten grains, with directions to repeat it three times a day.

On the 3rd "pain in the *head* and *arm* completely gone—*leg* no better."

On the 4th as on the 3rd, a blister to the nape of the neck, and eight cups over the lumbar spine.

On the 5th "says the cups almost immediately relieved the pain in his *leg.* He now feels perfectly well."

On account of the extreme rigour of the season, the patient was not discharged until the latter end of February, during which period he remained entirely free from disease.

CASE II.—Jane Black, aged sixteen, was admitted into the Hospital on the 9th of March 1831. About four weeks antecedently, she perceived pain, tumefaction, and a sense of numbness in her feet and ankles, which gradually deprived her of locomotion; and on the third or fourth day, confined her to bed. On the second day after the attack, her wrists and hands were similarly affected. In the course of a week her wrists, fingers and ankles, became flexed and rigid, feeling pain from every attempt to straighten them. Such was her condition when admitted. She states that she is of a costive habit, and had been amenorrhagic for two or three months before the appearance of rheumatism. The previous treatment consisted, as she said, of a blister to the *umbilical* region, and some powders and drops. On her admission,

Dr Norris applied six cups to the cervical, and six to the lumbar spine, which "took away entirely the pain." On the following day Dr Otto saw her, and recommended a continuance of the treatment, and accordingly four cups were applied to the upper and four to the lower part of the spine, with the effect of enabling her to extend her wrists, and to grasp, though imperfectly, with her hands.

On the 10th took Epsom salt.

On the 13th spine cupped as before and dose of magnesia directed. After the cupping to-day, she begins to observe a "pricking sensation, as if her feet and hands were asleep."

On the 16th cups as before.

On the 18th, find her free from pain and tumefaction, recovering gradually the use of her hands, experiencing no uneasiness on motion or pressure. She is unable to stand, because her feet "slide from under her;" but the attempt gives no pain. Besides the remedies already mentioned, soap liniment was applied twice a day to her wrists and ankles.

Remarks.—In this highly interesting case, the complication of rheumatic irritation with *numbness* and enfeebled condition of the extensors of the hand, and the congeneric flexors of the foot, amounting almost to paralysis, emphatically directs us to the centrally nervous origin of this disease.

CASE III.—William White, seaman, aged fifty-two, was admitted November 27th for rheumatism. He stated that he had an attack in the previous winter, which had confined him to bed for five months, and that the present affection had commenced with equal severity. On admission, his wrists and arms were tumid and painful, and he complained also of pain in the *lumbar* region and lower extremities. Cups were applied to his spine, and repeated at proper intervals, two or three times, without the use of any auxiliary remedies. The relief was almost complete, when in consequence of some accident, he was affected with fever and pain in the head, for which he was cupped and blistered at the nape of the neck, and a saline purgative given. Being relieved from the cephalic irritation, he began in a few days to complain again of pain in the feet and ankles which appeared hot and tumid. Cups having been applied to the base of the spine, entire exemption from pain ensued. The severity of the season prevented his discharge until the 26th of February; but for more than a month before, he had ceased to feel any other inconvenience, than a very slight soreness on the top of his feet, and that only when walking. That pain left him previous to his discharge. This case is reported by Dr Stewardson.

CASE IV.—William King, a seaman, was admitted for a surgical disease, for which he had used venesection and low diet, followed by balsam of copaiba and cubebs.

On the 24th, he was seized with severe rheumatic pain in his left side and shoulder. For this he was twice bled largely, and put under the use of sarsaparilla and nitrous powders, and afterwards of Dover's powders. A stimulant liniment was also applied to the affected part. Under this treatment he remained until the 6th of February, when the pain appeared to be fixed in both the side and shoulder, and he had not been benefited in any way by the remedies employed.

On the 7th of February, all other remedies being discarded, twelve cups were applied to the spine.

8th pain relieved. Cups to be applied.

11th patient states that the last cupping has almost entirely removed the pain from his shoulder, but has not benefited that of his side. Ordered eight cups to dorsal spine.

13th no change after last cupping. Cups to be again applied.

16th the pain in the shoulder left the patient soon after the application of the cups on the 13th and has not returned. As the pain in the side was confined at last to a small surface, and had been constant for some time, a few cups were applied immediately over it, with beneficial effect. Their repetition at length entirely removed it. This case is reported by Dr G. Norris.

Remarks.—The practical interest of this case consists in the total failure of the most judiciously selected remedies of the current practice, and the facility with which the disease, so obstinate before, began to yield to the very first application of cups. To those who still maintain the identity of the effect of general and topical depletion, this case presents a striking difficulty.

CASE V. William Brown, seaman, was admitted March 5, 1831, for rheumatism. Three months ago he was exposed at sea to great hardships in an open boat. On the day after he was picked up, he felt pain in his shoulders and elbows, which remained until after his arrival in port, and then suddenly attacked his lower extremities, while entire exemption from pain was experienced in his upper ones. On admission, he complained of pain in the whole course of his legs, but finds it particularly severe in his knees and ankles. The right angle is swollen,

L.   8 hot and very painful. Directed the application of ten cups to the *small of the back.*

March 6th. Is no better.—On examination, I found that the cups *had not been placed on the part as ordered,* but had been extended to the top of the spine. Therefore ordered another cupping to the loins.

7th. Was relieved by the cups for a time, but the pain has returned. Cups to be repeated.

8th. Has had very little pain since the last scarification. The tumefaction of the right ankle has disappeared, and the heat and pain have entirely gone from it.

On the 11th and 13th, in consequence of the reappearance of slight symptoms of the disease, cups were ordered. Their application in both instances afforded relief. Reported by Dr G. Norris.

Remarks.—In the case just recited, the attention of the reader is called to the fact, that the cups produced no relief whatever when applied over that part of the spine which did not transmit nerves to the seat of inflammation, thus verifying the important doctrine, that the most potent influence is exerted, when our *depletory* remedies are addressed as nearly as possible to the *disease exciting* agent.

Case VI. Thomas Gordon, a man of colour, a seaman aged thirty-four, was admitted on the 15th of February, for *rheumatic fever.* The pain is confined *chiefly to his limbs,* and his pulse, although excited, is not very active. Ordered ten cups to spine.

17th. No improvement. It is discovered that the cups had not been placed near the spine, but at a considerable distance on each side of it. Ordered ten cups to dorsal spine.

18th. The pain in his body and arms diminished, but no improvement observable in his lower extremities in consequence of which eight cups were applied to the *lumbar* portion of the spine. For a slight cough some mucilage was ordered. The patient was relieved by the last cupping and the pain *almost entirely* left him. For stiffness in his legs, a stimulant liniment was finally directed.

On the 1st of March, having been previously apparently cured, his disease suddenly returned. As he had along with other symptoms of fever, a strong and frequent pulse, sixteen ounces of blood were abstracted, and nitrous powders administered—but as on the following day, no abatement of the pain of the lower extremities appeared, and though the fever was reduced, eight cups were applied to the lumbar spinal region, which *entirely relieved* him.

On the 9th of March, he was discharged cured. After the last scarification, he used for stiffness and weakness of his joints a stimulating liniment. Reported by Dr Stewardson.

Remarks.—In this case several facts are worthy of notice. Twice the cups failed to relieve the *lower extremities*, once because they were not applied to any part of the spine, and once because they were placed on the *dorsal region.* The very first application to the lumbar region afforded the expected benefit. In the relapse, a large bleeding and nitrous powders sustained a total failure, while. a very moderate quantity of blood drawn from the lumbar region by cups, produced an immediate and final solution of the disease.

CASE VII. William Richardson, a seaman, was admitted on the 11th of February, for rheumatism. His attack commenced two weeks before, with pain in the dorsal region and occiput, followed by a *sense of numbness*, with pain in almost every part of his body. On admission his skin felt cold, his pulse was frequent, tongue slightly coated, and his bowels regular.

12th of February. Twelve cups were applied along the spine.

13th. *Has no pain;* slight numbness of the legs ; no appetite ; slightly vertiginous ; directed him an ounce of sulphate of magnesia.

14th. Nausea, for which ordered effervescing draught. For the numbness, directed soap liniment.

15th. No improvement; the numbness of his hands being especially disagreeable, a few cups were applied to the nape of his neck.

17th. *Find the patient free from pain and numbness.*

For an enlargement of the spleen, this patient remains in the hospital, but has not had any relapse. Reported by Dr Stewardson.

Remarks.—The most remarkable feature in this case is the concomitant numbness, and the greater difficulty of removing that than the pain, a fact which is not unfrequently observed in cases of rheumatism. The vertiginous affection too is interesting as significant of the irritation of central nervous masses.

CASE VIII. Rebecca Leshler, affected by rheumatism of two weeks duration, exhibited a swollen arm and shoulder, attended with pain and redness. She could elevate her arm only when firmly grasped by the hand of an assistant, when the motion became comparatively easy.

In the evening of the 5th of March, ten cups were applied so as to extend from the top of the neck downwards, immediately over the spine. On the following morning, the pain was gone, and on the subsequent day every vestige of redness and swelling disappeared. *No other treatment was used.* Reported by Dr Stewardson.

Although other cases might be cited in confirmation of the view here taken, I have not leisure at this time to digest and arrange them. At no very distant period I hope to be able to bring the subject more fully before the profession. I may observe in general, that, as far as I now recollect, only two cases of apparent rheumatism have in my hands, either in private practice or in the Pennsylvania Hospital, resisted the treatment recommended in this paper, and both of *them* were in reality *neuralgia,* and exhibited no traces of inflammation. One of them was an affection severely painful, located in the bottom of the heel, the other was gastric and intercostal.

The preference given to local depletion over other local measures, arose from the greater apparent success and promptness of its action, which scarcely left anything to be desired : but cases will occur in which other measures must be used, and in which perhaps all measures will fail. We are warranted, however, in declaring our conviction, that few failures will happen in thus treating *acute* rheumatism, and that success will diminish as passing through *chronic* rheumatism, we enter on the ground of neuralgia; a disease which sometimes spontaneously disappears, but is scarcely ever, in this city, cured by merely *medical* means. The art of the surgeon occasionally subdues it, and the physician often allays, but seldom removes it. Being paroxysmal, and often slumbering for weeks or months, it is not unfrequently mastered in appearance, though seldom cured in reality."

117

NOTE TO PAGE 47.

Criticism of Prof. Charcot* on the views of Dr Alison and Dr Brown-Séquard.

"It is manifestly to the arthritis of paraplegic patients, such as we have described it (*Arch. de Physiologie*, t. I), that the note of Dr Alison refers. It is a characteristic of the affection to remain confined to the paralysed limbs, and not to extend to the sound members. The affected joints are hot, swollen, and in some cases painful, either spontaneously or on movement made. The parts most frequently affected are the knee, elbow, wrist, hand, and foot. This form of arthritis seems to show itself chiefly in cases where the hemiplegia is consecutive on encephalitis or on brain softening. Two cases, selected from a number of others of the same kind, and cited as examples, deserve to be briefly recorded here:

CASE I.—A woman, aged 49 years, who had long enjoyed perfect health and had never suffered from any form of arthritic disease, was suddenly struck with hemiplegia; some days after, tumefaction and heat at the wrist of the paralysed side set in, and a little later on, the knee and foot of the same side became swollen and painful in their turn. There was no œdema. The paralysed limbs were rather rigid.

On *post-mortem* examination, partial softening of the brain was discovered. Each renal pelvis was filled with little calculi of uric acid.

CASE II.—A man, aged 54, house painter, who had experienced several attacks of gout, was struck with sudden hemiplegia. Soon after the wrist, the hand, and the foot, became hot and swollen. The paralysed limbs were rigid.

At the autopsy, the brain appeared softened, and a voluminous blood-clot was found in one of the lateral ventricles.

Dr Alison endeavoured to explain the occurrence of arthritis in the course of (hemiplegic) paralysis, by showing that "the healthy relation

* *Lectures on Diseases of the Nervous System.* New Sydenham Society, 1877, pp. 93—95.

between the living tissues and the materials of the blood was disturbed. Two morbid conditions gave rise to this disturbance, viz., a state of reduced vitality in the paralysed parts, and the presence of exciting and noxious agents in the blood. In proof of this various facts were referred to, and the author related two singular cases of the inflammatory red line of the gums following the use of mercury, in paralysis of one side of the face, being strictly confined to the paralysed side of the mouth. The paralysed parts were in fact more delicate tests of poisons than parts in a state of health. In proof of the presence of exciting agents in the blood the gouty diathesis of the second case and the lithic acid calculi in the pelvis of the kidney of the first case, were adduced."

We, in our turn, would point out that, most certainly, these cases are altogether exceptional, as regards the question at issue, for most frequently, as may be understood from a perusal of the cases published in our work (*Archives de Physiologie*, t. i), the arthritis supervenes in hemiplegic patients as a more or less direct consequence of the cerebral lesion, quite apart from all influence of gout, rheumatism, or other diathetic condition.

Hence, whilst acknowledging the accuracy of Dr Alison's clinical descriptions, I am unable to endorse the pathogenic theory which he has proposed. I am, however, far from denying that the articulations of paralysed members, in cases of hemiplegia of cerebral origin, may, as Dr Alison holds, be particularly disposed to become foci of elimination for other agents previously accumulated in the blood. I myself communicated to the Société de Biologie, at the time of its occurrence, a case in which this particular disposition was very prominent. A woman, aged about 40 years, had been suddenly struck with right hemiplegia, three years before her admission into my wards. The paralysed limbs were strongly contractured now and again, the several joints of these limbs, the knee especially and the foot, were the seats of tumefaction and pain. The patient, being aphasic, in a high degree, it was impossible to ascertain if she had been previously subject to gout or rheumatism.

At the autopsy, we found a vast ochreous cicatrix, the vestige of a focus of cerebral hæmorrhage, situated exterior to the extra-ventricular nucleus of the corpus striatum. In most of the articulations of the limbs on the right side, which had been hemiplegic, the diarthodial cartilages were incrusted towards their central parts with deposits of urate of soda, both crystallised and amorphous. The joints of the limbs, on the other side, presented no similar appearance. Some white striæ, which were found on microscopical and microchemical examination to be formed by urate of soda, were noticed in the kidneys.

It is undoubtedly most remarkable to find, in this case, that the gouty deposit forms exclusively in the joints of the paralysed members; but, I cannot too often repeat that facts of this kind are exceptional, and, in any case, they have nothing in common, from a pathogenic point of view, with the ordinary arthritis of hemiplegic patients ('Cas d'Hubert,' see Bourneville, *Études cliniques et thermometriques sur les maladies du système nerveux*, p. 58).

The merit is due to M. Brown-Séquard of having directed attention anew to the arthropathy of hemiplegic patients, and of having determined the organic cause, better than Dr Alison had done. He thus expresses himself in a lecture published in *The Lancet* (Lectures on the Mode and Origin of Symptoms of Diseases of the Brain, Lecture i, Part ii, *The Lancet*, July 13, 1861). After having admitted that the painful sensation, such as formication and prickling, which are experienced in the paralysed members, in consequence of a cerebral lesion, result generally from a direct irritation of the encephalic nerve-fibres, he adds:

'It is most important not to confound these sensations (which are referred sensations, like those taking place when the ulnar nerve has been injured at the elbow joint) with other and sometimes very painful sensations in the muscles or in the joints of paralysed limbs. These last sensations very rarely exist when the limbs are not moved, or when there is no pressure upon them; they appear at once, or are increased by any pressure or movement. They depend upon a subacute inflammation of the muscles or joints, which is often mistaken for a rheumatic affection. This subinflammation in paralysed limbs is often the result of an irritation of the vaso-motor or nutrition nerves of the encephalon.'

Before M. Brown Séquard, and before even Mr Scott Alison, many physicians had already remarked the arthritis of paralytic patients, but without bringing out the interest connected therewith. Consult R. Dann, *The Lancet*, t. II, p. 238, 1841. Durand-Fardel, *Maladies des Vieillards*, p. 131. Paris, 1854, Observation, Lemoine. Valleix, *Guide du Médecin Praticien*, t. IV, 1853, p. 514. Grisolle, *Pathologie Interne*, 2nd édition, t. II, p. 257."

NOTE TO PAGE 61.

On the synthesis of Xanthine, Hypoxanthine, Guanine, Theobromine and Caffeine.

By passing a stream of anhydrous hydrochloric acid gas, for some time, through anhydrous hydrocyanic acid, the latter being immersed in a freezing mixture, we obtain the hydrochloride of prussic acid, a white crystalline compound. On heating this with absolute alcohol the hydrochloride of methenylamidine or formamidine $CH\!\!\begin{smallmatrix}\diagup NH_2 . HCl \\ \diagdown NH\end{smallmatrix}$ is produced*.

If now we could effect a reaction between glycocine and methenylamidine similar to that which takes place between glycocine and urea to form hydantoic acid, we should have

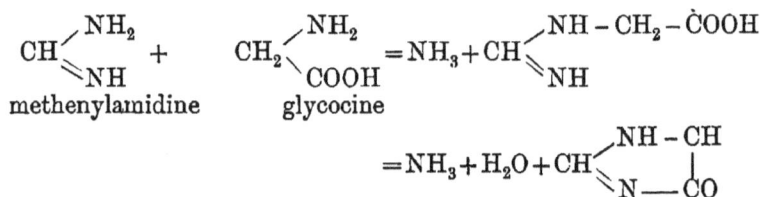

$$CH\!\!\begin{smallmatrix}\diagup NH_2 \\ \diagdown NH\end{smallmatrix} + CH_2\!\!\begin{smallmatrix}\diagup NH_2 \\ \diagdown COOH\end{smallmatrix} = NH_3 + CH\!\!\begin{smallmatrix}\diagup NH - CH_2 - \dot{C}OOH \\ \diagdown NH\end{smallmatrix}$$

methenylamidine · · · · glycocine

$$= NH_3 + H_2O + CH\!\!\begin{smallmatrix}\diagup NH - CH \\ \diagdown N\!-\!\!\!-\!\dot{C}O\end{smallmatrix}$$

Combining this last body with two molecules of urea, or with biuret we should obtain the formula for Xanthine—a combination similar to that by which uric acid is obtained from hydantoin and urea—

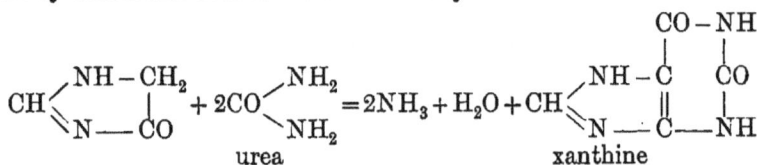

$$CH\!\!\begin{smallmatrix}\diagup NH - CH_2 \\ \diagdown N\!-\!\!\!-\!CO\end{smallmatrix} + 2CO\!\!\begin{smallmatrix}\diagup NH_2 \\ \diagdown NH_2\end{smallmatrix} = 2NH_3 + H_2O + CH\!\!\begin{smallmatrix}\diagup NH - C \\ \diagdown N\!-\!\!\!-\!C\end{smallmatrix}\!\!\begin{smallmatrix}CO - NH \\ CO \\ NH\end{smallmatrix}$$

urea · · · · · · · · · · xanthine

* For the mode of preparation of this substance, see Gautier, *Zeitsch. für Chemie*, 1867, s. 659, and Pinner, *Berichte der deutsch. chem. Gesell.* B. xvi. s. 375 und 1647.

If instead of taking two molecules of urea we combine one molecule of urea with one molecule of methenylamidine, ammonia being eliminated,

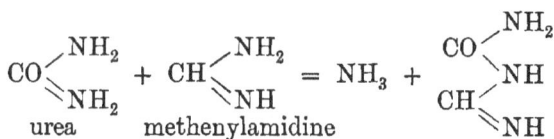

and then combine this product with the body $CH{<}^{NH-CH_2}_{N-CO}$ the formula would give us Hypoxanthine—

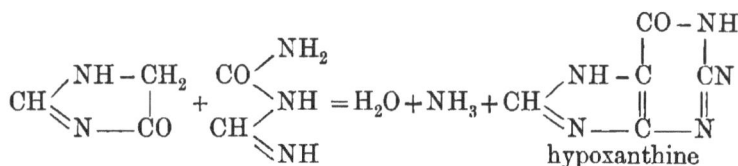

Similarly by combining $CH{<}^{NH-CH_2}_{N\ -CO}$ with guanidine $\Big($derived from biuret $NH{<}^{CONH_2}_{CONH_2} = CO_2 + NH{=}C{<}^{NH_2}_{NH_2}\Big)$ and urea, we should in the same way form Guanine—

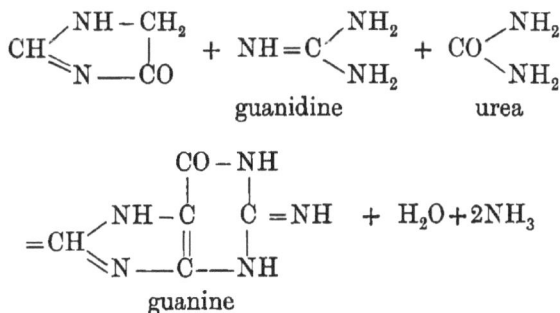

Substituting in this last equation, dimethyl carbamide for guanidine, we arrive at the formula for Theobromine—

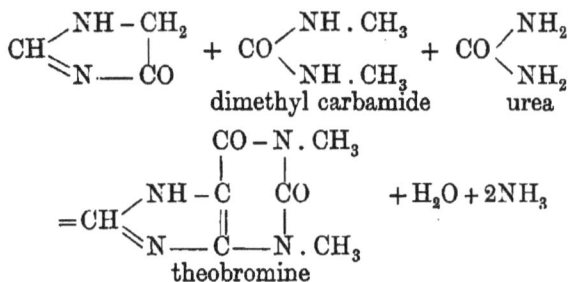

$$CH\Big<^{NH-CH_2}_{N-CO} \ + \ CO\Big<^{NH.CH_3}_{NH.CH_3} \ + \ CO\Big<^{NH_2}_{NH_2}$$

dimethyl carbamide   urea

$$=CH\Big<^{NH-\overset{CO-N.CH_3}{\underset{\parallel}{C}} \ \ \overset{}{\underset{\big|}{CO}}}_{N-C-N.CH_3} \ +H_2O+2NH_3$$

theobromine

And lastly, if instead of starting with methenylamidine, we take formimido ether $CH\Big<^{OC_2H_5}_{NH}$ (see Pinner, *Berichte*, XVI. s. 352 for its preparation) and act upon this with methylamine we have (Pinner, *loc. cit.*),

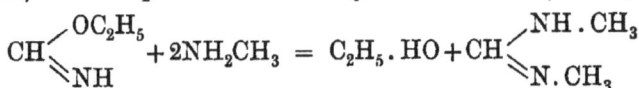

$$CH\Big<^{OC_2H_5}_{NH} +2NH_2CH_3 \ = \ C_2H_5.HO+CH\Big<^{NH.CH_3}_{N.CH_3}.$$

Combining the latter with glycocine, we have

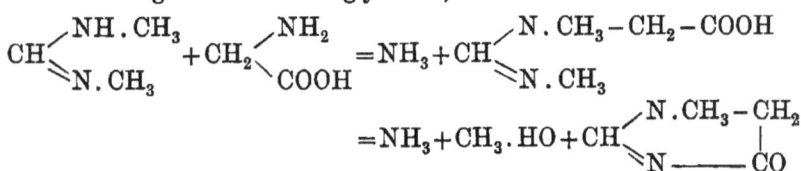

$$CH\Big<^{NH.CH_3}_{N.CH_3}+CH_2\Big<^{NH_2}_{COOH} =NH_3+CH\Big<^{N.CH_3-CH_2-COOH}_{N.CH_3}$$

$$=NH_3+CH_3.HO+CH\Big<^{N.CH_3-CH_2}_{N-CO}$$

which last combined with dimethyl carbamide and urea, as in the formation of theobromine, would produce Caffeine

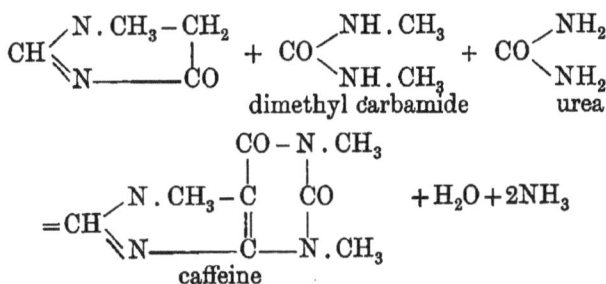

$$CH\Big<^{N.CH_3-CH_2}_{N-CO} \ + \ CO\Big<^{NH.CH_3}_{NH.CH_3} \ + \ CO\Big<^{NH_2}_{NH_2}$$

dimethyl carbamide   urea

$$=CH\Big<^{N.CH_3-\overset{CO-N.CH_3}{C} \ \ CO}_{N-C-N.CH_3} \ +H_2O+2NH_3$$

caffeine

During the past summer (1886) I have made a number of experiments with a view to obtain xanthine, &c. by the above methods; as yet, however, unsuccessfully. Results have been obtained which encourage me to continue the experiments, but the time required for them is considerable, and not easily found amidst the demands of active professional work.

CAMBRIDGE: PRINTED BY C. J. CLAY, M.A. & SONS, AT THE UNIVERSITY PRESS.